Farm Power Cyclopedia

A Book of Practical Information For Users of Engines For Farm, Shop and Household Purposes

by Berton Elliot

with an introduction by Roger Chambers

Self Reliance Books

Get more historic titles on animal and stock breeding, gardening and old fashioned skills by visiting us at:

http://selfreliancebooks.blogspot.com/

Introduction

I am pleased to present yet another title on Homesteading and Farm Life.

This volume is entitled "Farm Power Cyclopedia" and was published in 1915.

The work is in the Public Domain and is re-printed here in accordance with Federal Laws.

As with all reprinted books of this age that are intended to perfectly reproduce the original edition, considerable pains and effort had to be undertaken to correct fading and sometimes outright damage to existing proofs of this title. At times, this task is quite monumental, requiring an almost total "rebuilding" of some pages from digital proofs of multiple copies. Despite this, imperfections still sometimes exist in the final proof and may detract from the visual appearance of the text.

I hope you enjoy reading this book as much as I enjoyed making it available to readers again.

Roger Chambers

Fig. 11—C. E. COLBURN'S FARM AND STOCK BARN

1

Fig. 19—MR. LAWSON VALENTINE'S BARN, "HOUGHTON FARM," MOUNTAINVILLE, N. Y.

Fig. 69—INTERIOR OF COVERED CATTLE STALLS

Fig. 92—NOVA SCOTIA DAIRY AND FRUIT BARN

CONTENTS

CONTENTS

FARM POWER CYCLOPEDIA

INTRODUCTORY

Probably no one thing, since the dawn of civilization, has more completely revolutionized farm life in so short a period of time, as the application of gasoline power to farming requirements.

So important is the work involved, and so many and varied the applications of this form of power on the farm that there has been a demand for a standard, reliable work on Farm Power that will be broader and more comprehensive than the books heretofore published on the gasoline engine, and at the same time written in simple, non-technical language which can be readily understood by those without expert knowledge in the use of gasoline power.

In compiling this book, everything of an unimportant nature has been omitted, giving a condensed volume of information on the selection, use and application of gasoline power equipment, that will be of practical value as a constant reference book.

Convenient arrangement has been kept particularly in mind throughout, with the aim of permitting the quickest possible reference to any subject. The form of the book, also, is such that it may be easily carried in the pocket or tool kit, if desired.

The information contained in this book includes a digest of the most important statistics and data published in re-

9

cent years by the United States Government, and various State Agricultural Experiment Stations and Agricultural Colleges, as well as the knowledge and experience of many of the leading manufacturers of gas power equipment.

The author wishes to express his appreciation and thanks to the many who have so liberally extended assistance in the preparation of this work, and particularly The Goulds Co., The Deming Co. and The Fairbanks Morse Co. for material furnished on the subject of " Pumping," many of the tables given having been prepared by the engineers of these companies at great trouble and expense.

FARM POWER

A PRESENT DAY NECESSITY

The farm is the farmer's place of business.

The foresighted farmer knows that it is just as necessary for him to have up-to-date equipment as it is for the factory or business house.

The extensive use of power machinery on the farm has already made it practically impossible to compete in the agricultural markets of the world, by means of hand labor.

To make a success to-day, the tiller of the soil must multiply his own individual efforts. He must do more work in a given time, do it cheaper and better, and increase the yield per acre. He must spend less time in manual labor so that he can spend more time in intellectual work. The prosperous farmer of the future will be a *manager*, not a *laborer;* a good *business man,* as well as a *good farmer.*

GASOLINE POWER

THE IDEAL FARM POWER

The extensive use of gasoline power on the farm has not been a matter of chance.

It has been due to the fact that gasoline power more perfectly meets the peculiar conditions of farm usage than does any other.

The gasoline engine is always ready for use, can be started or stopped in an instant, uses fuel only while in operation and is adapted to a wider range of application on the farm than any other form of power. It requires less expert knowledge to operate than steam, and is less expensive and more generally available than electricity. It is compact, provides great ease of portability, and may be obtained in a closely graduated range of sizes, from one horse power for running churn or washing machine up to the largest size for the heaviest farm machinery. It is economical, safe and reliable and is not greatly affected by weather conditions.

PART I

THE GASOLINE ENGINE

T

T
mo
wh

line
un
th
gi
of
if

THE GASOLINE ENGINE

CHAPTER I

The " internal combustion engine," or, as it is more commonly called, the " gasoline " or " gas engine," is the unit which generates the power in all gasoline power equipment.

As certain underlying principles are common to all gasoline engines, regardless of design, model or size, a thorough understanding of these principles is the first essential in the use or application of gasoline power for any purpose.

Although at first thought the principles of a gasoline engine seem complex and difficult to understand, as a matter of fact, they are really quite simple and easily understood, if the theory of gas power generation is well fixed in mind.

CHAPTER II

OPERATING PRINCIPLES OF INTERNAL COMBUSTION MOTORS

The essential parts of every gasoline engine are one or more cylinders in which a sliding piston works back and forth; the piston being connected by a pin to a connecting rod, the other end of which is connected by another pin to a crank shaft which receives the back and forth motion of the piston and transmits it in revolving form to the point of application.

The chief fact to bear in mind in the operation of an internal combustion engine is that gas, when compressed and ignited, expands with great force.

The gas is drawn into the air tight cylinder and compressed.

At exactly the right time, it is ignited by an electric spark.

This produces a powerful heat expansion called an "explosion."

The explosion exerts pressure against the movable piston, causing it to move. The motion is transmitted to the crank shaft by the connecting rod.

The piston, through the momentum of a fly wheel, is brought back after each succeeding explosion to its initial position.

In producing each power impulse, four cardinal operations are required:

1. A charge of explosive gas or "mixture" (liquid gasoline mixed with air) is drawn into the cylinder or "combustion chamber."
2. It is compressed.
3. It is ignited.
4. The exploded or burnt mixture is ejected.

To this extent, the principle of all gasoline engines is identical. The method of performing these functions, however, differs considerably, and various auxiliaries are employed in their performance. In some engines the cylinders are placed in a horizontal position and in some, vertical, the movements and operations being the same in either case.

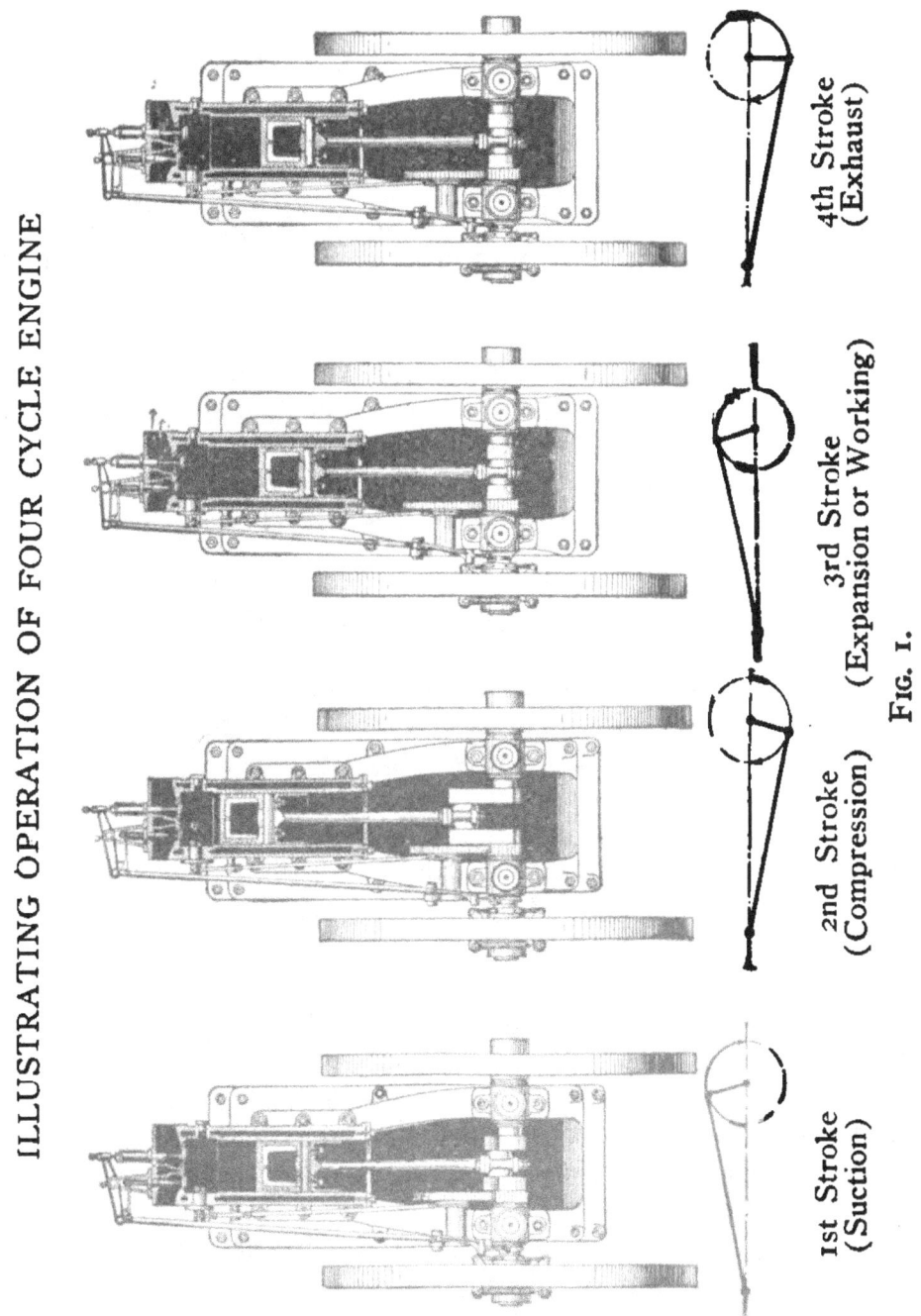

ILLUSTRATING OPERATION OF FOUR CYCLE ENGINE

4th Stroke (Exhaust)

3rd Stroke (Expansion or Working)

FIG. 1.

2nd Stroke (Compression)

1st Stroke (Suction)

CHAPTER III

FOUR CYCLE AND TWO CYCLE ENGINES

Gasoline engines may be divided into two general classes — four cycle and two cycle.

In the Four Cycle type, the complete cycle of operations required for each power impulse, viz: taking in the explosive mixture, compressing it, igniting it, and ejecting the exploded mixture, are accomplished in four strokes of the piston and two revolutions of the crank shaft.

In the two cycle engine, these operations are performed in two strokes of the piston and one revolution of the crank shaft. Two cycle engines are further divided into two and three port types. Four cycle engines are more commonly used at present for farm work.

The Four Cycle Engine

There are several types of four cycle engines, the design and general arrangement of parts being different, but the fundamental principles remaining the same.

Following the operation of a typical four cycle engine, refer to Fig. 1. Assume that the piston has just started its initial stroke, as shown in the first view at the left. On this stroke the forward movement of the piston causes a partial vacuum, or suction in the cylinder, the intake valve is drawn open and a charge of the mixture is drawn into the cylinder from the carburetor. This stroke is termed the suction stroke.

When the piston returns on its second stroke (see second view, Fig. 1), the explosive mixture is compressed into the head of cylinder. This stroke is termed the compression stroke.

Just before the piston reaches the highest point of compression, a contact is automatically made, allowing an electric current to pass from the batteries or magneto where it is generated to the interior of cylinder. This ignites the explosive mixture. The fraction of time which elapses between the time of ignition and complete combustion allows the piston to reach the end of its travel at just the time explosion has reached the highest pressure.

When the mixture in the combustion chamber is exploded and the high pressure between the head of the cylinder and the movable piston forces the piston forward (see third view, Fig. 1), this third stroke is called the power stroke.

Immediately before the piston reaches the end of the third stroke, the exhaust valve is opened by a cam, and the next backward stroke of piston (see fourth view, Fig. 1), expels the exploded or burnt mixture through the exhaust valve. This is called the exhaust stroke.

These operations are then repeated. Operations are performed in identically the same way in a multiple cylinder engine, each cylinder working individually.

The majority of farm gas engines are divided into two general classes; those in which the inlet valve is mechanically opened by cams, and those in which the inlet valve is automatically operated by suction of the piston on its charging stroke, as explained above. On all four cycle gas engines, the exhaust valves are operated by cams.

The principle of all four cycle engines is the same as shown in illustration (Fig. 1) but the location and arrangement of valves, etc., are different in various engines; inlet and exhaust valves are sometimes both on the same side of the cylinder, as in the L Head type, and sometimes on opposite sides, as in the T Head type.

The Two Cycle Engine

In two cycle engines, the cycle of operations required for each power impulse — taking in the mixture, compressing it, igniting it, and expelling the exploded charge — are performed with two strokes of the piston and one revolution of the crank shaft.

The piston on its upward stroke draws the mixture into the crank case from the carburetor. On the next downward stroke of piston, the charge in the crank case is forced up to the top of the cylinder through a bypass, where it is compressed by the upward stroke of piston, and ignited. The piston is then forced downward by the force of the explosion, the burnt gas or exhaust passing out through the port. These operations are then repeated.

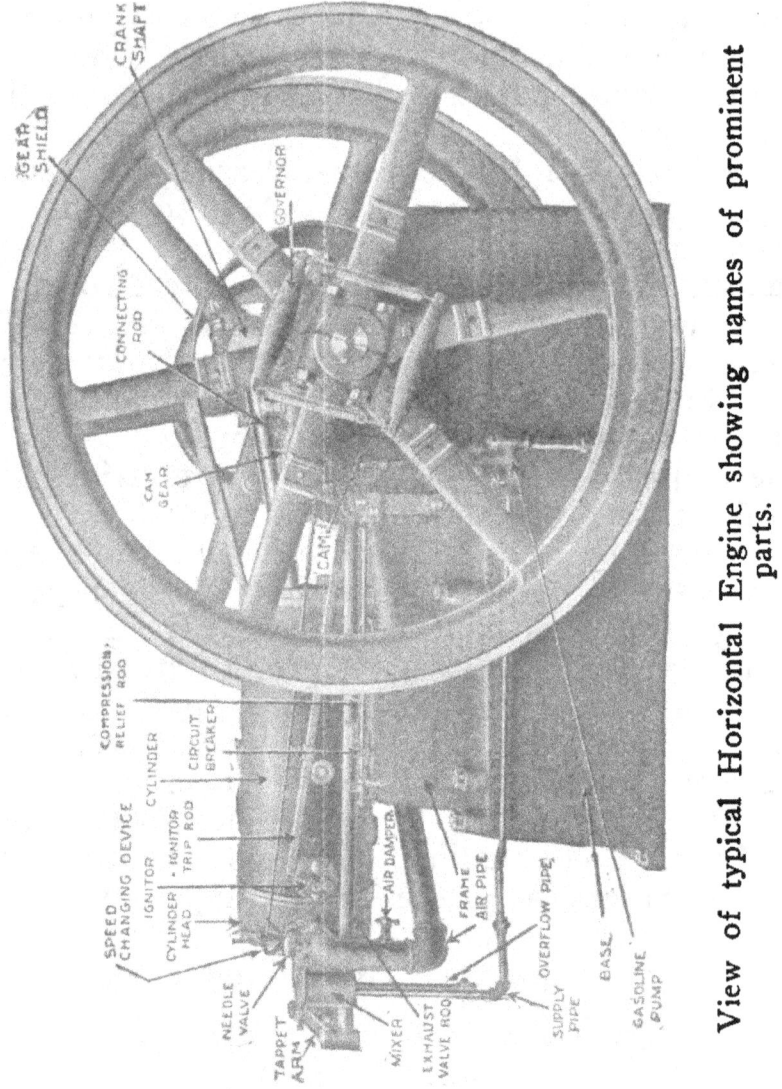

View of typical Horizontal Engine showing names of prominent parts.

Two cycle engines are made in both two port and three port types.

In two port engines, the fresh charge enters the crank case on the upward stroke of the piston, through a check valve which opens when the piston starts upward and closes automatically on its downward stroke.

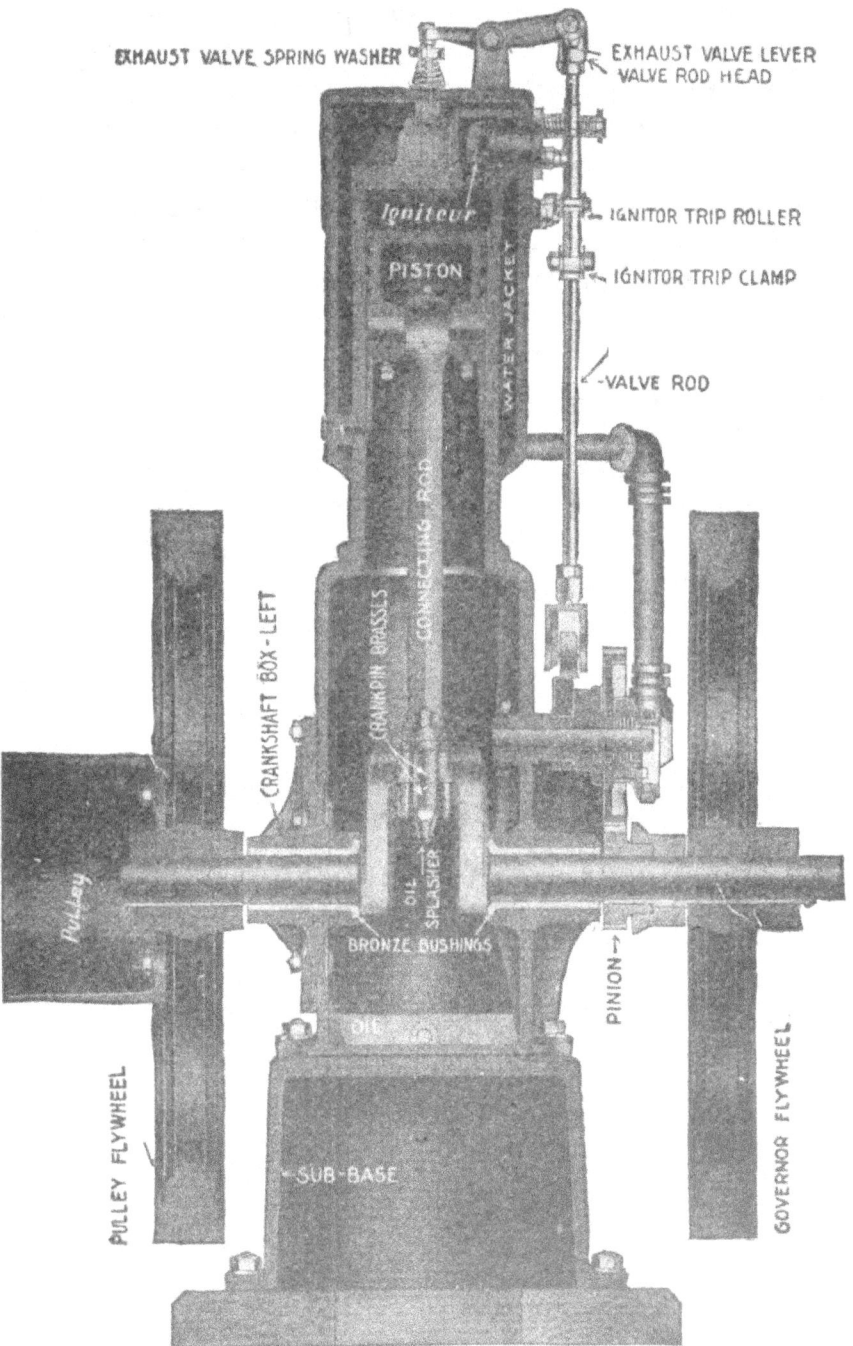

Sectional View of Vertical Engine.

It will be seen that in the two cycle engine, two strokes of the piston perform the same work for which four strokes of the piston are required in the four cycle engine, and theoretically it should develop double the power of a four cycle engine. This would be so, provided all conditions were ideal in each case. In two cycle engines, however, certain disadvantages combine to offset this. Among these are the fact that loss of power occurs through the waste of unexploded mixture through the exhaust port, and by imperfect cleaning out of dead or burnt charge before new charge enters, leaving part of the burnt charge in the cylinder to dilute the fresh charge; also loss of power through forcing the piston against the pressure in the crank case in driving the mixture via bypass into the cylinder head.

Another disadvantage of the two cycle engine is the fact that the piston must move back far enough to close the export before compression can begin; this shortens the effective compression stroke.

CHAPTER IV

PHYSICAL CONSTRUCTION OF THE GASOLINE ENGINE

Certain features of construction are considered standard in building gasoline engines, although considerable variation is made in the general arrangement and location of various parts by different designers.

The Combustion Chamber

The cylinder of the engine, which forms the chamber in which combustion occurs, is usually made of cast iron.

Ability to withstand the force of the explosion is one

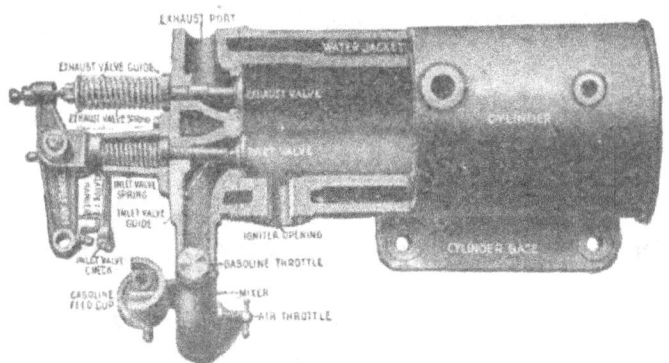

Sectional View of Cylinder.

of the requirements of the cylinder and is one of the important points to be considered in the selection of a gas engine. Gray iron is considered the best material at the present time. It should be uniform in texture. A quality of metal that will not expand evenly under excessive temperature is not satisfactory.

The cylinder must also be bored true, without the least variation in size and must form a perfect circle, the whole length. It must also be ground or reamed to a high degree of smoothness, to permit the piston to travel back and forth without undue friction or loss of compression, as well as to prevent the accumulation of carbon deposits, which adhere more readily to a rough surface.

Accessibility is another vital factor. The interior of the combustion chamber should be readily accessible for cleaning. The water jacket, if any, should also be easy of access for the removal of sediment or any substances which might clog the circulation of the cooling water. In four cycle engines, the valves should be easy of access for examination and for regrinding.

The Piston

The functions of the piston are such that greater care and accuracy are necessary in its construction than in any other part of the engine, except the cylinder.

It must be so accurately machined that it will fit snugly

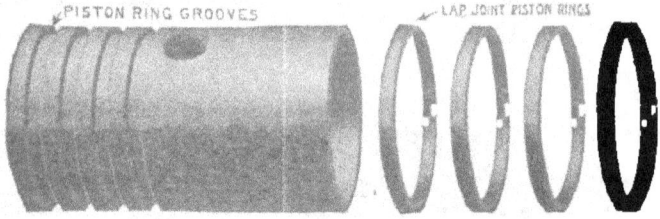

Piston and Lap Joint Piston Rings.

against the cylinder walls, so that practically none of the explosive mixture will leak past it when under compression, and at the same time permit of traveling back and forth within the cylinder with a minimum amount of friction.

As it would obviously be impossible to produce a solid, unyielding block of metal that would fulfill both these requirements with any degree of satisfaction, the method is employed of fitting rings into grooves cut around the outside of piston. The piston rings, which are split or cut on one side, are made of iron of a springy nature, giving them a tendency to expand and press lightly against the cylinder walls and at the same time yield sufficiently to minimize friction.

Piston rings must be tough enough to permit of spreading over the end of the piston without breaking and have spring enough to return to their original form.

From two to five rings are generally employed for compression purposes, with sometimes an additional ring provided to assist lubrication, by distributing the lubricating oil on the cylinder walls.

The Connecting Rod

The connecting rod, which connects the piston with the crank shaft, is an important factor of the gas engine, owing to the fact that the full power of the engine is thrown on this one comparatively small part.

Connecting Rod.

The connecting rod is attached to the piston by means of a piston pin and to the crank shaft by a crank pin. Both must be heavy and strong enough to withstand the sudden throwing on of the full load of the engine at every power impulse.

Connecting rods are generally drop forged steel, cast bronze, or machined from a solid billet of steel, or cast of malleable iron with adjustable bearings at both piston and crank shaft ends to take up the wear at these points.

Whether adjustable or not, bearings should be long enough to support the heavy load they are called upon to carry. They should be in perfect alignment and adjusted so as to be neither too loose nor too tight.

Bearings are usually lined with a bushing, babbitt, phosphor bronze or some metal alloy, so that in case of neglected or faulty lubrication, excessive wear, etc., the bushing may be replaced at small cost, without the necessity of a new bearing part at high cost.

The Crank Shaft

The crank shaft or main engine shaft receives the power

from the piston via the connecting rod and transmits it in rotating form to the point of application.

The crank shaft is another part of the engine that must be of very durable construction. The crank shaft is generally drop forged steel or cut from a solid steel billet, and must be ample in size to withstand the tremendous twisting strain to which it is subjected, and at the same time must be practically free from any tendency to crystallize from the effects of constant sudden strains.

The Base or Crank Case

The base or crank case is the casting which carries the crank shaft bearing and supports the cylinder.

In two cycle engines, where the crank case is utilized for compression, it must be made air tight, and should be as small as possible, yet permitting crank shaft and connecting rod to clear the walls in their travel.

In four cycle engines, the crank case serves as a bed or support for working parts.

The Fly Wheel

The purpose of the fly wheel is to aid the operation of the engine and make it run more evenly.

It gathers momentum from the first stroke of the piston, which momentum tends to equalize any unevenness arising from intermittent gas explosions and to keep the speed up when it would otherwise have a tendency to fall below normal.

Ordinarily speaking, the heavier the fly wheel the more efficiently it will do its work. However, if it is too heavy for the size of the engine, it will tend to slow down the engine and be too bulky for convenient operation as well as unsafe when running at high velocity, so that nicety of design is essential.

The Muffler

The function of the muffler is to deaden or muffle the noise caused by the exhaust of the engine. This should be done without creating back pressure on the exhaust which would reduce the power of the engine. Various types of mufflers are used for this purpose, the mechanical construction of which is not of essential interest to an engine owner or buyer.

In case the engine is set inside a building, an exhaust

Types of Mufflers.

pipe can be run to the outside and the muffler placed on the end of pipe outside of building, if desired. When this is done provision should be made for the protection of any inflammable material which might come in contact with the exhaust pipe.

The Governor

Practically all gas engines for farm use are provided with some sort of a governor for controlling or governing the speed of the engine.

There are two types of governors in general use — the throttling governor, and the hit-and-miss type.

The throttling governor operates directly on the gas admitted to the cylinder, feeding just the amount necessary to keep the engine running at a steady speed regardless of the load it is carrying.

The hit-and-miss

Type of "Hit-and-Miss" Governor operated from fly wheel of engine.

governor operates on the exhaust valve, and is so designed that when the engine is running above a certain speed, the exhaust valve is held open, allowing air to be drawn into the cylinder and preventing a charge entering cylinder. When the speed of engine drops below normal, the governor releases the exhaust valve, allowing a full charge to enter the cylinder.

In instances where constant steady running is necessary, such as electric lighting, flour mills, etc., the throttling governor is considered advisable, but for ordinary farm work, such as pumping, feed grinding, etc., the hit-and-miss is very satisfactory and is generally considered most economical in fuel consumption.

CHAPTER V

CARBURETION

The liquid gasoline or other fuels used in an internal combustion engine must be mixed with a certain amount of air to form the explosive mixture.

This process is called carburetion. Three types of devices are in common use for this purpose — carburetor, generator valve, and mixing valve.

The carburetor is the most highly developed of the three types, and consists essentially of a float feed, which automatically maintains a steady feed of fuel. There are many designs of float feed carburetors, but the general principle of operation is practically the same in all.

The carburetor is connected by a pipe with the fuel tank, which is generally located higher than the carburetor, although when convenient to do so it may be located at a lower point, and the liquid fuel forced up to the carburetor by air pressure.

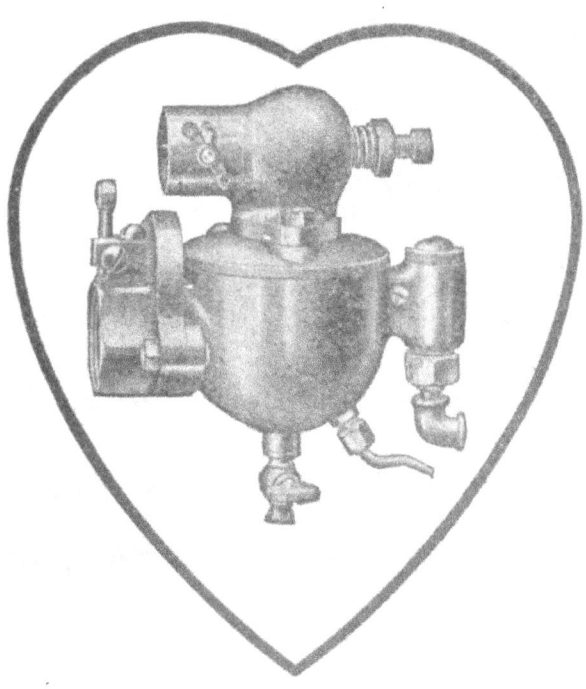

A type of Float Feed Carburetor.

The gasoline enters a part of the carburetor termed the float chamber, a uniform level of the gasoline is automatically maintained by means of a float of cork or other light

29

material, floating on the surface of the liquid fuel, to which is attached a long needle connecting with the fuel inlet valve. The float rises and falls with the level of the gasoline supply, opening and closing the valve.

The liquid gasoline passes from the float chamber,

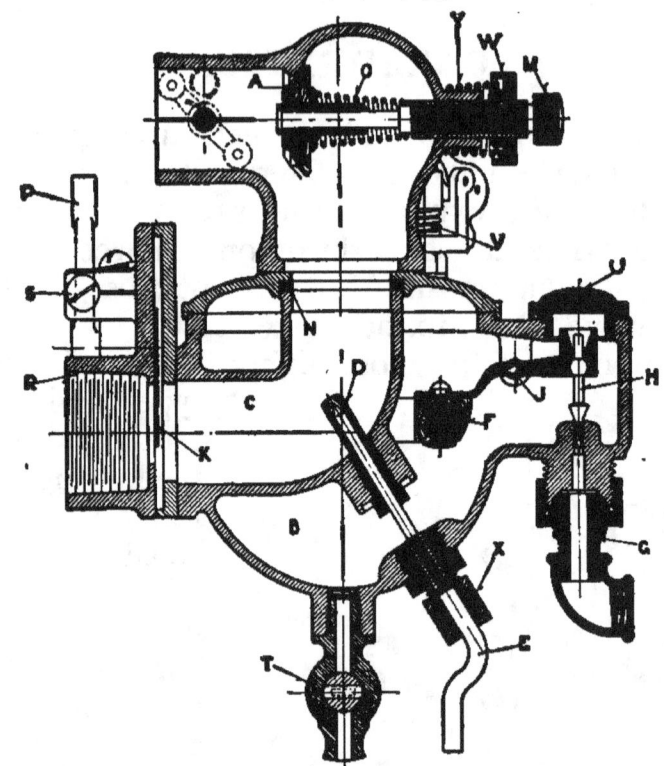

Showing operating principles of a type of Float Feed Carburetor.

through a spray nozzle into the upper mixing chamber in the form of a fine spray.

Air also enters the mixing chamber through an inlet provided for the purpose.

The air mixes with the gasoline spray, forming the explosive mixture, which passes from the carburetor into the engine cylinder through a throttle valve.

The amount of incoming air and gasoline, as well as the amount of mixture fed to the cylinder, can be regulated at will by the operator.

The less air to a given amount of liquid fuel the richer the mixture, and the more air the weaker the mixture.

The carburetor is a very delicate piece of mechanism, and inclined to be very "cranky" at times. When it is running well it should therefore be let severely alone, as the slightest change of any part is liable to cause serious trouble.

Carburetors are as a rule perfectly adjusted when sent out, and should not be meddled with by any one without expert knowledge, and never unless an accident or change of fuel makes adjustment necessary.

Generator Valve

The generator valve is much more simple in construction than the carburetor. It consists only of a spray nozzle, which governs the amount of gasoline fed to the mixture, and an air valve which governs the supply of air.

The generator valve when once set for a certain condition operates very satisfactorily, but has to be readjusted in either amount of fuel or air fed to the mixture, whenever the conditions change, consequently the generator valve is more sensitive than the carburetor, and, as it is not automatic in operation, requires constant attention.

Mixing Valve

The mixing valve is the simplest device for carburetion purposes. There is no set rule covering their construction as practically every manufacturer designs a device that is specially adapted to his particular engine. Some mixing valves are so extremely simple that they really provide for nothing more than a jet of gasoline sprayed into the air intake of the engine.

CHAPTER VI

IGNITION

In order to explode the charge or mixture within the combustion chamber of the engine, some systematic method of ignition must be used.

While various methods have been used during the various stages of development of the gas engine, the electric spark system is in almost universal use at present.

There are many types of electric spark ignition, but in all of them the same general principles are followed, producing electric current, intensifying it, conveying it to a point within the combustion chamber, and interrupting the current at intervals so that a spark will occur at just the proper time, when the explosive mixture is under the highest compression.

There are two general types of electric ignition in common use — the Jump Spark and the Make and Break systems.

The Jump Spark System

With the jump spark ignition system, the current is led through primary or low tension wires from batteries or magneto, where it is generated, to a spark coil where it is intensified or transformed from its primary or low tension stage to a secondary or high tension stage. The current then proceeds through the secondary or high tension wires from the spark coil to a spark plug which is screwed into the cylinder, and the points of which are within the combustion chamber.

The interruption of the current at regular intervals is brought about by a timer placed at a convenient point on the engine.

Jump spark ignition is becoming more extensively used on gas engines for farm use, owing to the fact that there are fewer mechanical parts subjected to wear and tear.

Another advantage of the jump spark system is the fact that none of the operating parts are permanently located inside the combustion chamber where they are hard to get at, as well as the fact that the spark plugs, where most of the wear generally occurs, may be replaced for one dollar or less.

The Make and Break System

With make and break ignition, the spark is caused by breaking the electrical circuit by pulling apart two metal points within the cylinder. These points, which have previously been in contact, are pulled apart by mechanical means at just the proper time, when the mixture is ready for ignition, and a spark is caused to bridge the gap between the two points.

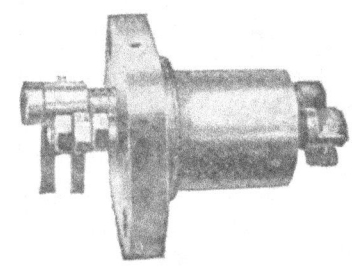

Ignitor, Make and Break Engine.

The make and break system is a later form than the jump spark, and is generally considered to furnish a slightly hotter spark. It is also less subject to short circuit by moisture.

A coil is used with the make and break method, which, however, does not transform the current, but simply receives the current from its source of supply and intensifies it when the circuit is broken.

Dry Batteries

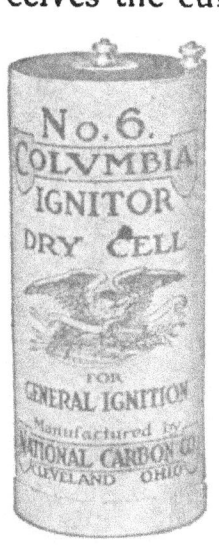

Dry Battery.

Dry batteries, or dry cells, which are one and the same thing, are in most extensive use as a source of electric current in farm gasoline engines, especially in the smaller sizes.

This is due to the fact that they are low in cost, very simple, easily connected and manipulated, and do not easily get out of order. Their great disadvantage is rapid deterioration, requiring frequent replacement.

A dry battery usually consists of two elements — one a carbon or negative, the other a zinc or positive — immersed in a moistened electrolyte paste. The carbon element is generally in the shape of a rod which stands upright in the center of a zinc container, in which the paste is packed. When properly connected by wires, a current of electricity will be set up between the carbon and zinc elements.

Four to six dry cells generally constitute a set or battery for gasoline engine work.

Wet Cells

Wet cells or batteries, as indicated by their name, use a liquid as an electrolyte, in which are immersed two pieces of metal, usually a zinc and a copper plate. Their operation, however, is not radically different from that of dry batteries, the plates being connected by wires and furnishing a continuous flow of electricity.

The chief advantage of wet cells is that the different parts are renewable, whereas in the case of dry cells it is necessary to procure a complete new battery.

In the selection of wet cells, one important thing to keep in mind is to see that the jar is substantial and so constructed that the chemical content will not creep over the edge and evaporate or corrode the wires and binding screws.

Storage Battery

Storage batteries are used more particularly where it is

A type of Storage Battery.

desired to have the flow of current independent of the running of the engine.

The storage battery does not generate current, but, as the name indicates, stores it for future use.

Storage batteries may be re-charged an indefinite number of times, the life of the battery being limited only by the life of its parts. It is necessary, of course, to take the battery to a re-charging station, which is not usually very convenient to a farm, unless a dynamo is installed for doing this.

A storage battery should always be set perfectly level, and so located that in case of leakage, the acid will do no injury to substances with which it may come in contact.

Magneto Ignition

Magneto ignition is the highest type of gas-engine ignition. Although not in such extensive use on the smaller sizes of farm engines, on account of its higher first cost, it is used very largely on tractors and other big outfits, requiring the highest efficiency equipment — and on any engine it is undoubtedly the cheapest in the long run.

A magneto, when once installed, requires practically no attention or expense for renewals. It will outwear any number of batteries.

Magnetos produce a hotter spark and are more positive than any other form of ignition.

A magneto generates electric current by employing the use of permanent magnets somewhat similar to the ordinary horse shoe magnet.

Generally speaking, the principle of a magneto is about as follows: A shaft wound with copper wire, and called an armature, is rotated across the ends of a magnet. This cuts through the magnetic influence that exists between the two ends or poles of the magnet, and sets up an electric current, which is collected by contact brushes, and can then be led to the spark coil (which on high tension magnetos is part of the magneto itself). The interior of the magneto should never be tampered with by any one who is not expert in electricity.

A magneto will last for years, in fact a good magneto will last as long as an engine. It should, however, be kept clean and properly oiled.

There are two types of magnetos — high tension and low tension.

Representative
low tension
Magneto.

Low Tension Magneto

The low tension magneto produces an alternating current, and having only a low tension wiring, must be used with a transformer or spark coil when used for jump spark ignition.

Batteries are not necessary with low tension magneto, but are often used, to make an easier starting outfit, as well as giving two independent ignition systems, so if one gives out the other can be employed.

Low tension magnetos are furnished in two types — one for use with make and break ignition — the other with jump spark ignition.

High Tension Magneto

The high tension magneto can be used only in connection with jump spark ignition. As it is provided with both a high tension and low tension wiring, the current is intensified without the use of the spark coil.

The spark plug points must be closer together in the high tension than in the low tension system.

Ordinarily speaking there are three types of high tension ignition.

In one type, magneto current only is employed.

A second type is the Dual System, wherein both magneto and batteries are employed. The engine is provided with two separate spark plugs, batteries being attached to one and magneto to the other. This system really represents two complete ignition outfits, each independent of the other, either

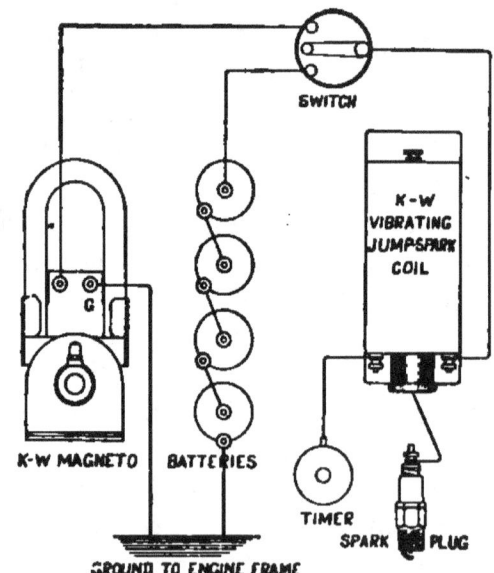

Simple wiring diagram for Single Cylinder Engine, with Low Tension Magneto and Battery.

one of which can be used as desired. If one system gets out of order the other can be used.

In a third type, magneto and batteries are used, so arranged that the engine can be operated on either batteries or magneto, through the same set of wires and same spark plug.

A representative High Tension Magneto.

Spark Coil

The current after being generated by batteries of low tension magneto is led to a spark coil.

Spark coils are of two kinds.

One form, used with the jump spark system of ignition, intensifies or transforms the current from its primary or low tension stage to a secondary or high tension stage.

The other form used with the make and break system, simply receives the current from the source of supply and acts as a reservoir, storing up current and discharging it on its journey when the circuit is broken.

No spark coil is used with high tension magnetos, the magneto itself performing the function of the spark coil.

The make and break coil consists of a magnetic core around which is wound hundreds of turns of fine insulated wire.

Spark coils for use with the jump spark system are given another winding of insulated wire called the secondary winding.

Type of Spark Plug.

Spark Plugs

The point of contact between the electric current and the explosive mixture is provided, in the jump spark system, by a device called the spark plug, which screws into the engine cylinder.

Spark plugs are of different design, but

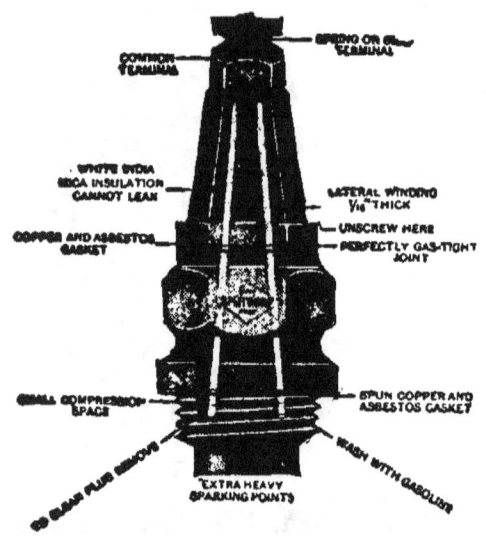

Illustrating typical Spark Plug construction.

all agree in principle to the following description of a typical plug:

A connection is provided at the top of the plug to which the wire from spark coil or magneto is attached. From this connection a wire runs the entire length of spark plug through an insulation of porcelain or mica. This wire terminates at the base of the spark plug in a sparking point inside the combustion chamber.

Rivetted into a shell of spark plug about $\frac{1}{32}$ to $\frac{1}{64}$ inch away from this point is another sparking point which is in contact with the metal engine cylinder, forming a circuit with the ground wire from the battery or magneto.

When the current is interrupted it jumps from the sparking point connecting with spark coil or magneto to the sparking point connecting to the ground wire, causing a spark to jump the gap between the two points, which ignites the mixture.

The Timer

Ignition timing is not an electrical process, but is controlled by a mechanical device called a timer.

The timer is most generally operated from the cam shaft of the engine. Timers are of different designs, but generally consist of a spool containing alternate contact segments and insulation. The spool fits over end of timer drive shaft and revolves with the shaft.

There are two general classes of timers in most common use — the

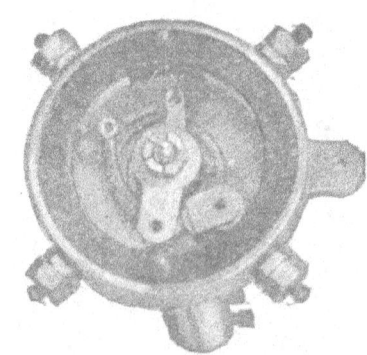

A type of Roller Timer.

roller contact type, in which the contact is made by a brush roller held against contact spool by a spring — and the slide contact, in which the revolving contact spool slides over a straight surface of brushes held against it by a spring.

The timer is generally operated from the crank shaft of the engine by two gears meshing together.

CHAPTER VII

COOLING

The intense heat within the combustion chamber and cylinder of a gasoline engine necessitates some method of cooling the cylinder walls enough to prevent the lubricating oil being burned.

Two methods are in general use — air cooling and water cooling. Oil is also used to some extent, having the advantages of lack of evaporation, and permitting use in localities where water is scarce.

Air cooling is very extensively used on farm engines of the smaller sizes, and is especially adapted for use where the engine is only run at full capacity for short periods of time.

The principle generally employed is to expose as large a surface as possible to the air. This is done by casting fins or ribs on the cylinder and head. A fan is also often employed to force air against the cylinder.

Water cooling is also very largely used on small farm engines, and is used almost altogether on large, heavy outfits, and on engines that have to work under a constant full load.

The water cooled system consists essentially of surrounding the cylinder and cylinder head with water. The walls are cast double and form a water jacket.

The water is kept in constant circulation, being forced from a tank or other source of supply. On farm engines, the water is usually returned to the tank after passing through the jackets, the same water being used over and over again.

Two methods of circulating the cooling water are in general use — pumping and the thermo-syphon or natural heat system.

Various types of pumps are used, the most commonly

employed being the rotary and plunger types, operated from the crank shaft of engine, and the centrifugal pump.

The plunger pump consists essentially of a barrel in which a plunger is tighly fitted, which is operated by a cam on engine crank shaft or gear shaft. The suction and discharge valves are so arranged. that on the forward stroke of the plunger, the vacuum created draws the water through the lower or suction valve into the pump barrel. On the backward stroke of the plunger, the force of the water closes the suction valve, and passes upward through the discharge valve into the water channels of the engine. To

Centrifugal type Water Circulating Pump.

keep pump running, all joints should be kept tight and free from air leaks, and packing around pump plunger should be perfectly tight, so that no water can leak out.

The rotary pump consists of two spur wheel gears, operated by either gears or belt from the crank shaft or gear shaft. The meshing of the two gears between themselves creates a suction which draws the water into the pump and forces it into the water channel. Rotary pumps should only be used where clear water is available, as sand or sediment will soon cut gears, destroying their suction qualities.

The centrifugal pump is very similar to the rotary pump in action, but consists of a single wheel or fan, which in itself draws the water up to its blades and forces it on throughout the circulating system.

In the thermo-syphon system, an inlet pipe runs from

tank to the lower part of water jacket, the tank being so located that the water outlet of tank will be higher than the inlet of cylinder. An outlet pipe runs from the top of cylinder jacket back to the tank. The water runs from the tank to lower part of jacket, and as water when heated rises, the hot water passes out through the outlet pipe in top of jacket back to the tank to be used again.

The gravity feed of the thermo-syphon system is the simplest and least expensive, and is entirely satisfactory unless some obstruction should occur that gravity would not have sufficient force to overcome.

Plunger type Water Circulating Pump.

Hopper cooling is the most extensively used method on the smaller sized farm engines. It consists essentially of a hopper or receptacle surrounding the cylinder, the hotter water going to the top of the hopper, where a surplus of water is maintained.

Water cooling, while more positive than air cooling is at a disadvantage in the winter time, owing to liability of freezing unless watched carefully. A great many engines are now claimed to be free from danger of freezing, and it is true that great progress in this respect has been made during the last few years, nevertheless it should be considered part of the regular attention of a gasoline engine in winter to drain water jacket and pipes thoroughly after using and to protect from the cold as much as possible.

Anti-freezing mixtures are often advocated, but great care should be exercised in their use, as they are frequently injurious to the engine, and often a mixture that gives satisfactory results in one engine will be disastrous in another, under different conditions.

CHAPTER VIII

LUBRICATION

Lubrication has more to do with the efficient operation of a good piece of machinery than almost any other factor, and is the greatest agent for long life and minimum depreciation.

There is no machine in which lubrication is of so much importance as in a gasoline engine. This is due to the excessive temperature within the cylinder space, which is equal to that of the combustion of the mixture, whereas even in a steam engine the cylinder is never hotter than the steam, and the particles of water in the steam aid in reducing friction.

The method of lubrication used, therefore, must be especially adapted to high temperature. Ordinary machine oil will not answer for the lubrication of gasoline engine cylinders, and neither will an oil suitable for steam engines.

Cheap oil should be avoided, the quantity of oil being of no advantage if it is poor in quality.

A good cylinder oil should be of heavy consistency, with a heavy flashing point, so that it will not burn readily and pass out with the exhaust without lubricating the cylinder walls and piston, and not carbonize or cake.

Particular attention should be given to obtaining an oil adapted to the particular make of engine in use. It is generally well to follow the recommendation of the engine manufacturer as to the particular brand of oil he has found most satisfactory to use.

The system of transmitting the lubricant to the points of friction is equally as important as the lubricant itself; it must be constant and dependable, and fed at just the proper speed. A number of lubricating systems are in general use to-day, some of which are here described.

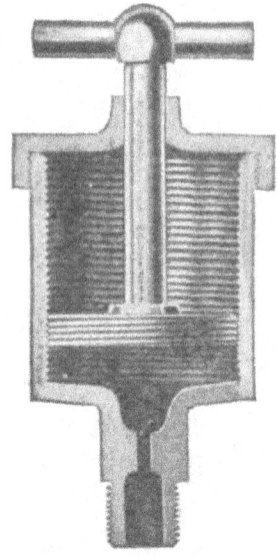

Compression
Grease Cup.

The Gravity System

The gravity system, by means of grease cups, sight feed oilers, etc., which was in extensive use a few years ago, is the simplest method of lubrication. It consists simply of grease cups or oilers placed in an elevated location directly over each point to be lubricated. The two objections to this system are that it is very inconvenient to operate, every cup having to be turned on and off separately — and that it is liable to become clogged more easily than where force is used. The liability to forget turning oiler on when starting is also an objection, resulting in scored cylinders and pistons, injured crank shaft, burned out bearings, etc., as is the liability to forget turning it off when through, resulting in wasting lubricant as well as causing engine trouble. Grease or oil cups also waste lubricant, as the feeds are usually set for maximum service and left that way regardless of speed and lubricating requirements.

The Splash System

This system, generally speaking, utilizes a tight reservoir or oil container in the crank case or at some convenient location surrounding the crank shaft. The moving connecting rod dips into the oil, splashing it onto the crank shaft and carrying it up into the cylinder.

There are various modifications of this system, all of which are similar in general operation. The splash system is often used in addition to a force feed or mechanical oiling system.

Force Feed Systems

With force feed systems, the lubricant is forced through separate feed pipes to

Sight Feed Oiler.

the points of friction, automatically with the operation of the engine. The lubricating system starts and stops with the engine, and regulates the rate of feed to conform to engine speed. A force feed oiling system relieves the operator of annoyance and worry. Except for an occasional filling, it does the necessary thinking and remembering automatically. If of proper design and construction, it affords three essential advantages — 1, it avoids any possibility of forgetting to turn oiler on when starting engine — 2, it avoids possibility of forgetting to turn oiler off when engine is stopped, preventing waste of oil — 3, it has sufficient force to overcome any obstructions which may become lodged in pipes.

The Mechanical Oiler

The mechanical oiler consists of a pump or series of pumps, the plungers of which are operated by the mechanism of the engine. The amount of oil fed to the engine varies as the speed of the engine, affording uniform lubrication at all times.

A type of Mechanical Oiler.

Mechanical oilers can, with expert mechanical knowledge, be installed on most engines by the owner.

Fuel Lubrication

A recent method of lubrication, which is now advocated to a considerable extent, particularly on two cycle engines, consists of mixing the lubricant with the fuel. This method has the advantage of doing away with the necessity of special lubricating equipment, and also insures the oil getting to all parts of the cylinder. A disadvantage claimed for it, however, is that the mixture of lubricating oil with the fuel lowers the specific gravity of the fuel, making it more difficult to vaporize. As this method only takes care of lubricating the cylinder and its auxiliaries, where used it is necessary to lubricate bearings, crank case parts, etc., by grease cups, or some other method.

Where fuel lubrication is employed, from a pint to a quart to each five gallons of gasoline is generally used.

CHAPTER IX

INSTALLATION AND ADJUSTMENT OF A GASOLINE ENGINE

A gas engine when shipped from the factory is more or less completely assembled ready to start. There is, however, usually a certain amount of adjustment and general installation work to be done before engine is ready to crank.

There is such a variety of engine designs and so many methods of installation, depending upon the requirements, as well as the personal ideas of various engine experts, that it is impossible to give a set of instructions that will apply in all cases. There are, however, a number of general principles that hold good in all instances, or which can easily be modified to fit some particular conditions. These will be given. Wherever an engine is purchased new from the factory, a set of instructions covering this particular make of engine is almost invariably sent out by the manufacturers. If this is lacking the makers should be written to about it as most likely it was overlooked.

Upon arrival of the engine at the freight depot, it should be looked over very carefully to see that it has not been broken or damaged in transit, or that none of the parts are missing. If not in perfect condition, the railroad agent should be requested to note the deficiencies on the receipt before you sign it.

Assuming that the engine is in perfect condition, it should be hauled home and uncrated.

If engine is of the stationary type, a foundation should be provided for it. The better the foundation, the better results the engine will give. Even a sewing machine runs best on a solid floor. A good cement foundation is usually best. Directions for laying a cement foundation will be

47

found further along in another Chapter on " Stationary Engines." If it is not desired to go to the trouble of laying a cement foundation, a very satisfactory foundation can be made by attaching engine bolts and lag screws to two 4 x 4 or 6 x 6 timbers, which should be in contact with the ground their full length. Stakes can be driven to take the pull of the belt, if desired or found necessary.

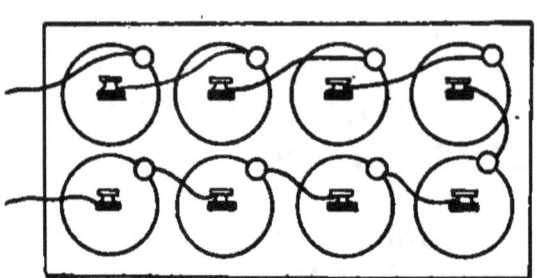

Series Connections.

If engine is to be installed in connection with line shafting, general instructions for installation will be found in a later Chapter on " Shafting, Pulleys, and Bearings," also in Chapter on " Power House Construction."

If engine is to be moved about wherever it is desired to drive machinery, it should be mounted on skids or wheeled truck of some sort. If the outfit is not purchased mounted in this way, skids can be easily constructed at home.

Engines are generally wired up, but as this is not always done, or connections become loosened, or some different ignition wiring system is desired, some general information on this subject will be given.

Care should be taken to see that batteries are wired correctly; that is, when batteries are connected in series, carbon element should be attached to zinc element, not carbon to carbon, nor zinc to zinc. In case of connecting up in multiple, all carbon elements should be connected together on one circuit, all zinc elements on another. When connected, see that there are two free binding posts, one carbon and one zinc, and be sure all connections are perfectly tight.

In wiring batteries, secondary or high tension wires should not be allowed to get any closer than one-half inch to low tension cable or to any metal except terminals of spark plug and coil, and particular care should be taken to

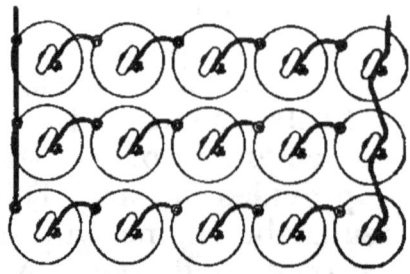

Multiple-Series Connections.

guard it from being saturated with oil or water.

Spark coil should always be located so that all wiring will be short as possible.

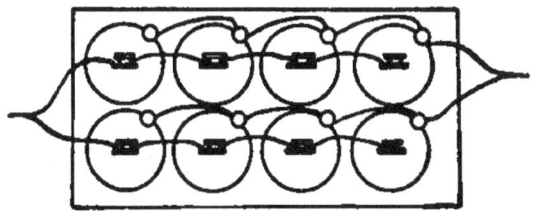

Multiple Connections.

The entire electrical outfit should be kept dry, as a drop of water at any of several places will put it out of commission. If engine is left out of doors, care should be taken that covering is provided, so that no rain or dew can get to the spark plug, cables, switch coil and battery box.

After all installations are completed, go over engine carefully to see that all connections are tight. See that oiling and cooling systems are in good working order, and that all screws, bolts, and nuts are tight. After you have assured yourself on all these points your engine is ready to run.

CHAPTER X

HOW TO OPERATE A GASOLINE ENGINE

The first thing to do in starting a gasoline engine is to be sure that you have plenty of fuel and lubricating oil.

In filling gasoline tank, fuel should be strained through chamois skin, or one of the various types of strainers especially made for this purpose. It is very little trouble and the entrance of dirt is liable to interfere with the adjustment of the carburetor, generator valve or mixing valve.

Open water inlet, if any.

Turn on fuel both at tank and engine.

If the weather is at all cold prime cylinders by introducing about a teaspoonful of gasoline in priming cups provided for this purpose.

See that lubricating system is adjusted to feed the proper amount of oil.

Retard the timer to starting point.

Close switch.

Turn fly wheel over.

If engine does not start with a few turns, something is probably wrong. Don't keep on cranking, but look for the reason.

First thing, see if fuel tank is filled and fuel supply turned on. An empty fuel tank and failure to turn on the fuel supply have doubtless been the cause of non-starting more often than anything else.

If fuel supply is O. K. failure to start may be due to a great many reasons, chief of which will be explained in the following chapter on "Engine Trouble": the most common cause of non-starting, after lack of fuel, however, being ignition trouble, batteries, switch, magneto, etc., should be first examined to see if anything out of order can be readily detected.

In stopping the engine, first shut off fuel and lubricating oil, then open switch.

50

On water cooled engines, the plug provided at lowest point of water jacket should always be removed after use in freezing weather, and water drained out thoroughly.

After using the engine it should be wiped off well with waste or a rag. (Never do this while engine is running, as waste may be caught in running parts and cause a serious accident.)

If engine is to remain idle for any length of time, it pays to wipe over all unpainted parts with waste dipped in oil to prevent rust.

Starting in Cold Weather

A gasoline engine is always harder to start in cold weather.

It is often necessary to prime the cylinders in winter, as previously explained.

Warming the cylinders will in many cases be found a remedy. With water cooled engines, a simple and often efficient way is to pour a bucket or two of boiling water into the water jacket or hopper, which can be easily done with the later type of engines.

See that everything is ready before pouring in water so that it will not become cooled.

Points to be Observed in the Operation of an Engine

After engine has been started, there are a number of things that should be watched.

Cylinders should be observed to see that they do not overheat. With hopper cooled engines it is permissible to let the water boil when engine is working at full load. If cylinder overheats, first see if spark is retarded. If this is not the case, the trouble is probably due to using too much gasoline or improper circulation of cooling water.

Lubrication should be watched closely at all times, no matter what system is used. More attention is needed when the machine is new, before the oilers are in proper adjustment and the bearings worked in. Too much oil is bad, just as much as too little. The exhaust should be watched to see if smoking — black smoke indicates too much gasoline being fed to the carburetor, generator valve or mixing valve, and blue smoke indicates too much oil being fed to cylinders through the lubricator. The exhaust should show just the slightest tinge of blue vapor.

Cautions

Do not screw spark plugs into cylinders too tight. This is one of the principal causes of breakage of spark plug insulators. Screw in just enough to avoid leakage. Care should also be taken not to allow water to splash on spark plug porcelain, while engine is running, as this is almost sure to crack it.

In small engines where started with a crank always pull up on handle — never push down, or it may kick and result in serious injury to the operator.

Always wipe off any surplus oil, which is liable to collect dirt and work into the bearings — but never do it while the engine is running.

In make and break engines occasionally examine igniter points, as they gradually burn and wear short until they will not reach far enough to make a circuit.

Keep batteries dry. They weaken rapidly if allowed to get wet or too damp.

Examine oiling system to see that no oil holes or pipes are clogged or leaking.

Don't forget to turn off fuel and oil when through running engine. If engine has suction feed it is not necessary to turn off fuel.

Don't start water cooled engines without seeing that water is turned on.

Don't start with spark too far advanced. It might result in a broken arm.

It is a good plan to keep extra batteries and spark plugs on hand.

Never look for leaks with a lighted match.

Keep engine in as solid and firm a position as possible.

Use good oil.

In using stick or wire to measure fuel in tank, be sure it is clean, otherwise it may be the means of dirt getting into the carburetor.

CHAPTER XI

HOW TO LOCATE TROUBLE

The gasoline engine with its countless advantages on the farm is not without one disadvantage —" engine trouble."

The more thoroughly the principles of the gas engine are understood and the more experience the operator has with an engine, the less trouble he will have, at the same time there will always be some little things that can only be determined by experiment until the difficulty is located. Reference to the usual causes of trouble which follow, and a little patience will, in the majority of cases, enable the operator to locate and remedy the faults with comparative despatch.

One thing to be kept in mind is that the same effect often arises from different causes — or in other words different causes may produce the same symptoms.

There is no question that at least half of the trouble with gasoline engines is due to the operator's forgetfulness. Something that should have been attended to is overlooked.

Where to Look First

When anything goes wrong the first thing to do is to go over the simple but essential things which must always be attended to in starting an engine.

If engine does not start when it is cranked:

First, see that fuel tank is not empty and that fuel is turned on.

Second, see that switch is closed.

See that battery connections have not become loose.

See that fuel supply line is not obstructed so that it does not feed properly.

See that air vent in fuel tank is not obstructed, creating a vacuum in tank and preventing fuel from flowing.

Obstruction of fuel supply can be detected by cranking the engine with open switch so that ignition system is cut out. If fuel supply is all right (and valves in four cycle engine are working properly) there will be a strong odor of gasoline in the exhaust that can be readily detected by any one standing back of the exhaust while engine is being cranked.

If tests do not show simple omissions or negligence of this kind, a systematic search for the trouble should be begun.

Ignition Troubles

The majority of actual trouble cases (outside of the simple cases of oversight) are due to faulty ignition, so this is the best place to look first, unless the symptoms clearly point to some other form of trouble.

With battery ignition, the first thing to do is to see that the wires are tightly connected to binding posts.

If no wires are found loose remove the spark plug and lay it on top of cylinder with wires attached, using care that wire terminals do not touch cylinder. If a good bright spark occurs between the sparking points it shows there is nothing wrong with the ignition system, unless it should be in the timing. If spark is of a very pale color or no spark occurs, remove the wire from the spark plug, hold the engine wire close to the engine frame, but not quite touching it. Have some one crank the engine. If a good spark occurs it shows at once that something is the matter with the spark plug, and the best way out of the difficulty is to put in a new one. In making this test be sure not to hold end of wire too far away from engine frame (as a rule not over $\frac{1}{8}$ inch), otherwise serious injury may result. If no spark occurs, or it is weak, the difficulty is probably between the spark plug end of the wire and the battery, and short circuit or broken wire should be looked for.

If there is no indication of a broken wire, examine the batteries. See that the terminals are free from grease and the contact good. See that wires are properly connected, none of them broken and no short circuits.

If everything looks all right disconnect wiring and test each battery separately, with an ammeter, or by the less certain method of bringing the ends of two wires attached to battery terminals together and pulling them apart.

If a battery is found dead or weak it should be replaced at once.

If batteries are in good condition and no evidence of broken wiring or short circuit is found, the next place to look is the spark coil.

First, see that wiring connections to the spark coil are all right. If so, test the coil for short circuit or broken down insulation.

Unfasten the secondary cable from the terminal. Attach in its place a short piece of wire which will reach to within about $\frac{1}{8}$ inch of a primary post. If no spark occurs, examine the platinum contacts between vibrator and screw to see if clean, then tighten and loosen a few times and test again. Then examine timer spring to see if weak or broken. If still no spark, it is a pretty sure sign that coil is burned out, in which case it should be sent back to the manufacturers.

If a good fat spark jumps the gap, it shows that there is no trouble here. If an intermittent spark occurs, the platinum points should be examined for pitting. If this is found, points may be smoothed down with an oil stone or fine file.

Where magneto ignition is used, the field magnets may need remagnetizing or the armature winding may be burned out. If this is the case magneto must be sent back to the manufacturers.

Fuel Trouble

Dirt or water in the carbureting device will cause trouble. Most carburetors can be cleaned by flushing or removing dirt through a drain pipe provided for this purpose. To avoid trouble of this kind always see that fuel tank is clean and dry before filling, and strain the fuel through chamois or a gasoline strainer.

Flooding is another form of fuel trouble that is very prevalent. This is nothing more or less than flooding the cylinder with too much gasoline. Often this is the result of over anxiety or unfamiliarity on the part of those un-accustomed to using a gas engine, but this is not always the case.

Flooding is sometimes caused by dirt under the fuel valve. When this is the case, it can often be stopped by flushing the carburetor. If carburetor is fitted with a drain open this and hold the needle valve open for a short while.

Flooding may also be caused by valve not seating prop-erly. This can be often overcome by tapping the needle gently. This valve should never be ground and if in poor condition the carburetor should be returned to the fac-tory.

Another cause of flooding may often be found in the float. Cork floats sometimes get soaked with fuel, so that they do not press valve tightly shut. All cork floats are covered with shellac, but this sometimes becomes worn off or cracked. A fresh coat will put them in good condition again, or a new float can be obtained at small cost. Metal floats sometimes become punctured and the hollow inside fills with fuel, weighting it down. A leak can be detected by shaking float and listening for fuel inside.

Flooding of a carburetor while engine is running is evi-denced by black smoke from the exhaust (if smoke is blue or white an excess of lubricating oil is indicated). If the engine is flooded it will be hard to start. To remedy, screw down the needle valve of carburetor generator or mixing valve until it seats, open the compression cocks on the engine and turn flywheel several times. This will gen-erally crank the excess gasoline out of the cylinder so that the engine can be started.

The proper fuel level in most carburetors is usually just below the tip of spray nozzle — about $\frac{1}{32}$ to $\frac{1}{16}$ inch.

Operating Troubles

In addition to the various troubles which prevent an engine from starting at all, are many operating troubles which tend toward unsatisfactory results and inefficient operation, some of the chief of which follow:

Hard Starting

Often times the engine can be started — showing that the

fuel and ignition systems are at least in fairly good condition — but at the same time be very hard to start.

Among the causes which contribute to make an engine hard to start are the following:

Air leak between the carburetor and cylinder.

The trouble may also be in spark coil — vibrators stuck, points fused, dirty or roughened, insulation damaged, etc.

Often an engine will start harder after it has been running awhile and stopped than when cold. This may be the result of heat thinning lubricating oil so that compression is lost past the piston rings. A small amount of thicker oil on the piston will usually alleviate this trouble.

When an engine has not been run for some time it is liable to start hard. This may be due to piston rings sticking. To overcome this condition put a generous supply of kerosene in the cylinder, and rock the flywheel back and forth a few times; then put in a little lubricating oil. The kerosene cuts the old gummed oil that has stuck the rings and the lubricating oil assures compression.

Hard starting in winter may be caused by too heavy a grade of fuel. As light a grade as can be secured should be used in cold weather, as high test liquids evaporate much more freely than low test ones at a low temperature. Hard starting in cold weather can often be avoided if engine is cranked two or three times with spark turned off and the throttle thrown wide open, then close the throttle two thirds and close the switch.

In four cycle engines the reason for hard starting is often found in the valves. Improper adjustment of the lift rod may hold them open, or stem may be gummed up, causing it to stick in its guide. The stem also may be bent.

When the adjustment of lift rod is correct, a small space, equal about to the thickness of a calling card, should be left between the end of valve stem and lift rod, when the valve is seated and the lift rod at its lowest point. If stem is gummed a generous application of kerosene will take care of the trouble. A bent stem, however, will generally necessitate removal of the valve to straighten it.

Lack of Power

Sometimes an engine will start all right, or perhaps a little hard, run perfectly on no load or light load, but when the spark is advanced and the full load thrown on will act "cranky" and slow down or refuse to pick up the load.

This may be due to several causes.

The mixture may be too rich. The remedy is to regulate the mixture according to the type of carburetor fitted to the engine. In cold weather more gasoline is needed to start them, but after engine is warm the feed should be reduced. This is also the case with low grade fuel.

Weak batteries or magneto or a leak in the coil may be the reason for trouble, there being enough energy to cause the spark to occur at low speeds with a rich mixture, but not enough to produce a spark at high compression with a lean mixture.

The most likely place, however, to look for this class of trouble is in the compression. There are several causes of weak compression.

Broken or worn piston rings is a frequent cause. They are generally the result of using too little oil. A continuous clicking or squeaking noise in the cylinder is usually a sign of worn or loose rings. This condition should be remedied at once, or the cylinder walls may become scored, meaning great expense for a new cylinder or regrinding and new piston and rings fitted.

A frequent cause of poor compression in four cycle engines is leaky valves. A little dirt or some foreign substance, possibly a grain of metal or emery from a former grinding, may have lodged on the seat of the exhaust or inlet valve, preventing it from seating properly. Valves also may become pitted. The exhaust valves get very hot while engine is running and burned oil may stick on the seat and be pounded into the metal, pitting valve and seat so they will not fit closely, thus allowing compression to get by. If valves are pitted they must be taken out and reground.

Other causes for lack of compression are a crack in the piston head, flaw or sand hole in the casting, rings losing their elasticity, cylinder walls worn out of round so that piston does not fit closely all the way around.

If a hissing sound is heard when flywheel is cranked slowly it indicates that compression is leaking into the air. This usually comes from leaking compression cocks, spark plugs or valves. A more certain method to determine if compression cocks or spark plugs are leaking is to put oil around them and crank the engine. If the oil bubbles it shows the leak.

Back Firing

This very annoying and somewhat common fault is due to the fact that at least part of the explosion occurs outside of the combustion chamber and is indicated by loud, sharp reports like a gun in the intake pipe or carburetor.

Back firing is usually due to some derangement of the carburetion system, and sometimes by ignition derangement.

Back firing is usually the result of ignition of the mixture in the inlet pipe by a flame in the cylinder left over from the previous explosion after the inlet valve opens.

Too weak a mixture may be the cause. The adjustment to feed less air to the amount of gasoline fed will usually correct the trouble. If this does not do it, however, look for derangement of the carburetion system — fuel tank may be nearly empty, fuel pipe may be clogged with dirt or sediment, spray nozzle, inlet valve, or air valve may be out of adjustment, or valve spring may be weak or broken.

A mixture that is too rich to burn well in the cylinder will also cause backfiring, in which case try feeding more air or less gasoline to the mixture.

If the trouble is not located in the carburetion system, look for it in the ignition system. A late or retarded spark will cause back firing.

In the case of continuous backfiring try feeding a little more fuel. If this does not stop it look for carbon deposits in the cylinder or leaking inlet valve.

Misfiring

Missing explosions is another form of trouble that is quite prevalent. It may be caused by almost any of the derangements that cause backfiring.

Missing on Low Speed

Frequently an engine operates perfectly on high speed and misses explosions on low speed.

This is not necessarily the fault of the carbureting system, and the best place to look first is for a leak in the inlet pipe between the carburetor and engine. This is sometimes indicated by a whistling sound. A good way to detect leaks of this kind is to place a gasoline soaked rag around the joints of the pipe one at a time. If engine runs better with fuel soaked rag around joint it shows this is where the trouble is.

Another very frequent reason for this form of trouble, in four cycle engines, is the valves. Examine them to see if they are clogged or need regrinding. Also look into the timing of valves, as often one of them may hold open when it should be closed.

Missing on One Cylinder

When one cylinder of a multiple cylinder engine is found to be missing explosions, the first thing to do is to locate which cylinder is missing. This can usually be done by short circuiting the spark plugs one at a time by holding a screw driver against both the base and terminal plug at the same time. Listening to the exhaust while doing this will show which one is missing. (Be sure to use a wooden handle screw driver and hold by the handle.)

The most frequent causes of misfiring in one cylinder are leaky or fouled spark plug, valve not seating properly and carbon deposits in the cylinder.

Irregular Missing

Other causes of missed explosions of a more or less irregular nature are: poor contact at commutator and ground wire, wires broken or saturated with oil or water, broken connections, spark coil vibrator out of adjustment, imperfect insulation of wiring system, weak batteries, etc.

In adjusting spark coil vibrator, be careful not to break springs if vibrators are stuck. If platinum points are burned out or pitted, smooth them off with a very fine file or fine emery cloth.

In make and break engines, trouble is sometimes caused by an accumulation of carbon at the sparking points. If this is found to be the cause, remove igniter and dress up points.

Pre-Ignition

Pre-ignition, as the name indicates, means the ignition of the charge before the proper time. The result is that the explosion is hurled against the piston while still on its inward stroke. The momentum of the piston, aided by the flywheel, is generally sufficient to pull the piston past center, so that the engine does not necessarily stop running. However the waste power is enormous, and serious damage will be done if neglected. Pre-ignition is indicated by pounding, and is sometimes mistaken for loose or worn bearings.

Pre-ignition is often caused by small pieces of carbon deposit or soot which have been heated to a red or white heat.

Faulty cooling is sometimes the cause, as is also imperfect lubrication of the piston.

It may also be due to advancing the spark too far.

If engine continues to run after switch has been opened it is a sure indication of pre-ignition.

Engine Stopping

When an engine stops suddenly, it is nearly always due to ignition trouble or stoppage of the fuel supply.

If engine slows down when under load, especially if accompanied by backfiring or missed explosions it is very likely to be due to improper mixture or weak batteries, loose connections or soot on the spark plug.

If engine slows down but does not backfire or miss explosions, it is a good plan to look for faulty lubrication, an overheated piston or a hot box.

If engine keeps slowing down and picking up again, it often comes from a loose wire or weak batteries, and in four cycle engines from weak exhaust valve spring.

If engine dies down slowly, the same as when switch is opened, the trouble is likely to be a broken wire, short circuit, dirty spark plug, broken timer or slippage of timer, throwing spark out of time.

If engine speeds up perceptibly and backfires just before it stops, it is quite probable that fuel tank is empty or fuel pipes clogged, or there may be water in the fuel.

If engine appears to be working hard, the same as when heavily overloaded, and finally stops, it is a pretty sure sign of friction or binding. Examine oiling system, and look for a hot box, or misalignment.

Knocking

Knocking or pounding should have prompt attention, as it is likely to indicate something serious out of order.

Knocking may result from a worn or broken piston pin, crank pin, connecting rod bearing or loose flywheel, also piping or rods that strike some part of the engine.

Knocking is usually purely mechanical; however, it sometimes comes from spark being set too early, pre-ignition of charge, faulty lubrication, and carbon deposits.

It is a good plan to try changing the time of spark before looking anywhere else for the trouble.

A knock can be located by placing a pencil between the teeth, stopping the ears and placing other end of pencil on various parts of the engine.

Never run an engine any longer than can be helped after knocking has been noticed.

Overheating.

A gas engine should run constantly and indefinitely without overheating, and where a tendency to overheat is shown something is wrong.

Imperfect cooling and imperfect lubrication are obviously the most usual causes of overheating.

Before going over the entire cooling and lubricating systems there are some other points contributing to overheating which it would be well to look into.

Engine may be heavily overloaded. Principal cause of overloading is incorrectly proportioned pulleys.

The pulley on the driven machine should be of such size as will allow the engine to run at normal speed.

Spark may be improperly timed.

Exhaust valve may not lift properly.

Mixture may be too rich or too weak.

In some cases overheating is due to faulty construction, in which case it probably cannot be overcome. However, if engine has ever been run without overheating, faulty construction is eliminated as a possible cause.

Spark Timing

Although no one method of spark timing will apply absolutely in all cases, the following is a simple method that is effective in most cases, under average conditions. It is to advance the spark control lever about one third while piston is at top of cylinder. Then move the contact point of the timing device in the direction it turns when engine is in operation, until vibrator begins to buzz. At this point fasten the contact spool to the driving shaft. This allows the spark to be advanced two thirds and retarded one third from the inner center position, and in cranking the engine the spark control lever should be less than one third advanced.

Valve Timing

Unless the valves on a four cycle engine open and close at just the proper time with reference to the position of piston and crankshaft, trouble will result.

If the exhaust valves do not open soon enough or wide enough the engine will lack speed. If they close too soon spark plugs will foul and engine will overheat and lack power.

The exhaust valve should be opened approximately 35° before the crankshaft reaches the end of power stroke, and close not sooner than end of exhaust stroke, and not later than 20° past dead center. This permits as much of the exploded charge as possible to be forced from the cylinder and next charge to be pure and fresh.

The inlet valve should not open before exhaust valve closes, as otherwise some of the burnt charge remains in the cylinder to mix with the fresh charge.

Valve Grinding

The perfect condition of valves in four cycle engines is

such an important matter that every owner and operator should know how to grind valves.

First, remove valves and place cotton waste or rags in the opening in cylinder where valves seat.

Place a little grinding material on valve and drop it into place, without replacing spring.

Then rotate the valve on its seat, with a screw driver or by some other means. Do not keep turning valve in one direction, but turn it back and forth. At short intervals, say every four or five turns, lift valve from seat to eliminate any possibility of cutting rings on the valve or seat.

Keep up the grinding until valve and seat show a bright, clean surface all around.

Before replacing valve permanently, be sure that every particle of grinding material is removed from both valve and seat, as it will tend to interfere with the operation of the valves and if any of the material finds its way to the cylinder it is liable to score piston and cylinder walls.

There are a number of preparations sold for grinding. Care however should be taken to use a good reliable material. Flour emery mixed with oil to form a paste is used extensively and so is ground glass.

CHAPTER XII

OVERHAULING AND REPAIRING THE GAS ENGINE

Like any other piece of machinery, a gas engine will give better results and last longer if kept in good condition and overhauled occasionally.

Obviously no hard and fast rules can be given for overhauling and disassembling an engine, as there are so many designs, differing in details of construction. However, a general method of procedure can be given which will enable any one to keep his engine in good repair by applying ordinary mechanical principles.

In overhauling engine, it is well to have it under cover and mounted on a block or low bench where convenient to get at.

First strip engine of carbureting and lubricating mechanism, piping, connections, etc.

Each part, when taken down, should be thoroughly cleaned with gasoline or kerosene, and marked with a prick punch or tagged so you will know how they go together. Bright metal parts should be covered with vaseline to prevent rust or tarnishing.

Right here, it should be emphasized that the various parts should be put away so you will be sure to know where they go, and where to find them when you want them. As it will assist if the various pieces of each particular part of engine are kept together, a good plan is to get a few boxes of different sizes, each to accommodate the parts of some particular division of the engine.

Engine should be drained of oil, fuel and water.

After connections, piping, etc., are detached, remove the cylinder or, if engine has a detachable cylinder head, removal of head is usually sufficient, which renders the cleaning of combustion chamber comparatively easy.

If necessary to remove cylinder, loosen the bolts holding it to crank case. It may or may not be necessary to loosen and remove piston, depending upon the design of engine.

If engine has received good average care, the cylinder will be found fairly smooth its entire length, and the old oil, etc., can be cleaned out by drenching with kerosene (do not use gasoline, which tends to deteriorate the surface of the iron). If there is a slight accumulation of carbon deposit, the kerosene will loosen it, so that it can be readily removed. Great care should be taken in scraping an engine cylinder not to injure the surface of cylinder or to scratch the walls.

The piston should next have attention, especially the rings. If rings are bright and smooth, it can be pretty well assumed that they do not permit compression to leak past, but if they have sooty streaks on them, it is almost sure that they leak.

Ordinarily it is not necessary to remove piston rings in order to clean them. Where necessary to remove rings to correct a roughened edge, great care should be used by those who are inexperienced not to break them in taking them off.

If rings are removed, they should be marked so that each ring will be returned to its own particular groove, and grooves cleaned thoroughly when rings are off.

In replacing rings, a good plan is to use three or four thin metal strips for slides and expanders until the ring is slipped over its groove, when strips may be removed. In replacing rings, be sure that the slits in two successive rings do not come in a direct line with each other.

If the pistons are removed from cylinders, in multiple cylinder engines, they should be marked so that each will be returned to the proper cylinder.

Piston pin should be examined and tightened if loose, or replaced if worn.

Connecting rod also should be examined. On some engines a hand hole is provided in the crank case through which the connecting rod is accessible. If there is no hand hole, cylinder must be removed, and if it is found necessary to remove crank shaft, the gears must be marked so they may be put back in the correct position. In going over connecting rod, if any bearing has become worn or has an unusual amount of play, it should be remedied. Some engines are made with adjustable bearings which provide for taking up wear, on others new bushings are furnished by manufacturers, and on some it is necessary to send it to a machine shop to be re-babbitted.

In four cycle engines, the valves may be more or less sooted, in which case it is necessary to grind them in, directions for which will be found in previous chapter on "Engine Trouble."

If valves are in extremely bad condition, it is well to remove the worst of the deposit with a fine file.

Timer should not be disassembled until the original setting as furnished by the factory has been marked.

Bearings that are fouled only in spots can generally be scraped down to work satisfactorily.

Crank shaft should be examined for uneven wear or misalignment.

Wiring terminals should be well cleaned with gasoline.

Spark coils should be cleaned, but with a care not to get insulation wet.

Spark plugs should be removed, examined, cleaned and adjusted.

Fuel tank should be cleaned out thoroughly, and examination made for rust or scale.

The above are the general points to follow in overhauling an engine, about as complete as it is possible to give without following some particular design of engine. In addition, a sharp lookout should be kept for anything that is out of whack — leaks, obstructions, dirt, broken parts, stripped threads, loose connections, etc.

When everything has been overhauled and cleaned, the engine is ready to be assembled. Before starting to re-assemble, care should be taken that no dirt or obstructions have gotten into any of the open parts, and before any part is replaced, it should be seen to be scrupulously clean.

Extreme care should be taken to put everything back just the way it was in the first place.

In tightening bearings, great care should be used. Each

side should be tightened slightly, working from side to side, instead of tightening one side complete before starting the other side.

Great care should be used in setting all nuts, especially small ones, not to strip the threads.

If engine is of multiple cylinder type, a straight edge should be placed across faces of exhaust flange bosses to see that they line up.

After engine is erected, a supply of oil should be placed in the oiling system, and the flywheel cranked a few times to furnish all bearings and wearing surfaces with a liberal supply of oil. About a cupful of oil should also be placed on each piston, where it should be allowed to stand until engine is ready for use again. Before engine is started the first time, this oil should be thoroughly drained off, and a little kerosene poured in so that any oil that might have become gummed and sticky is washed away.

CHAPTER XIII

ADVANTAGES OF GAS ENGINE OVER STEAM ENGINE

The fact that gasoline engines have, in a few short years, come into so much more extensive use on the farm than steam engines, despite the many years the latter have been on the market in a perfected state, is proof positive that they are much better adapted to farm requirements.

Simplicity and safety are two important reasons. Imagine a man turning a steam engine over to his wife, daughter or small son. It would be too complicated for them and the risk would be too great, but with the gasoline engine, the farmer has no misgiving after knowledge of operation is acquired.

Economy is another factor. It is commonly estimated that one pint of gasoline per horse power will operate an engine an hour — a slight cost in comparison with the work done. The economy, however, lies not only in the fact that the fuel cost while engine is operating is less than that of a steam engine, but also in the fact that when the engine is not working, no fuel is consumed, nor is any preparation required before starting engine. With the steam engine, the fire must be started one or two hours before commencing work to provide steam pressure, and often it is advisable to keep it up over night.

A gasoline engine can be started instantly and it is ready to take a full load from the start.

In the case of a steam engine it also usually requires the entire time of one man to run the engine and fire the boiler, while with the gasoline engine no attention is necessary after starting engine, adjusting lubricators, etc.

CHAPTER XIV

GAS ENGINE FUELS

Natural gas is the most convenient fuel to use in internal combustion engines, but as natural gas is practically unavailable on the farm and also impracticable for use in tractors and portable outfits, liquid fuel is universally used for farm purposes.

Gasoline is by far the most extensively used, as it is most ideal, all conditions considered. Gasoline is highly volatile and lends itself most readily and efficiently to vaporization.

Kerosene is now coming into somewhat extensive use and is generally considered more economical than gasoline. It, however, has the objections of being harder to vaporize as well as accumulating carbon deposits more rapidly, necessitating more frequent cleaning of cylinders, and causing a greater amount of engine trouble. The difficulties in using kerosene have to a degree been overcome in the design of some engines, but unless an engine is specially adapted to burning kerosene fuel, or has kerosene attachments, its use is liable to be the source of excessive trouble.

Distillate is another fuel that is in considerable use in some sections, particularly on the Pacific coast. It is in reality a very low grade of kerosene oil, and although low in cost must be handled with great care to avoid carbon deposits and serious operating trouble.

Benzine and alcohol are also sometimes used, benzine being highly satisfactory, but generally so high in cost as to be prohibitive.

The use of gasoline should be attended with great care, but if proper precautions are taken and ordinary common sense employed, it is not at all dangerous. If gasoline is kept within the receptacles intended, care is taken to avoid leakage or spilling, and lighted matches and open flames not permitted to be near it, no damage can be done.

As gasoline evaporates with readiness, it should be kept enclosed tightly, otherwise the fuel will disappear with startling rapidity.

In case a fire should start, however, it should not be attempted to extinguish it with water, which has little effect on burning gasoline; in fact generally tends to spread it. Sand and sawdust are good extinguishers, and a few buckets of either one of these materials kept nearby is a precaution well worth while.

CHAPTER XV

GASOLINE AND OIL STORAGE

The storage of fuel and oil on the farm is an important matter.

A representative Underground Gasoline Storage System being used to fill an Automobile.

First, as a matter of economy. Unless great care is used much waste will occur — the same as with hay, corn, etc., unless properly stored. This is more necessary with gasoline than with oil, for the reason that its highly evaporative nature will make it extremely expensive if improperly stored.

Second — proper storage will prevent deterioration of quality, which not only means the use of more fuel and oil, but decreases the efficiency of the engine, as well as increasing trouble in operating the engine.

Third — in the case of gasoline, careful storage is important for the reason of safety.

As the first two reasons are self evident, and common to both gasoline and lubricating oil, while the feature of safety is peculiar to gasoline only, the storage of *gasoline* will be taken up in particular for the time being.

The first essential in proper storage is an air tight tank.

The next essential is that it be protected from the access of children, the curious and the careless.

Rather than keeping it unprotected and free to general access, it should be kept under lock and key, in a shed.

Barrels and tanks carelessly stored above ground, even under lock and key, however, are not ideal, as the door may be left open, and the fluctuation in temperature also tends to reduce the quality of gasoline.

The best method of storing gasoline is underground. At a distance of two feet or so beneath the surface, the temperature is cool and varies little all the year round, preventing gasoline deterioration.

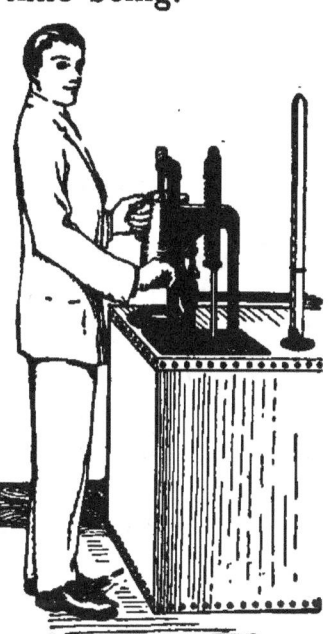

A modern Above-ground Tank Storage System.

For most efficient results, one of the special gasoline and oil storage systems on the market should be used. The cost is not excessive, and the value is very great. The following description of one of the representative underground systems of this kind will serve to show how these modern systems operate in the main.

A heavy, reinforced tank, as nearly air tight as possible, is placed underground at a convenient point, and connected by means of a pipe with a pump at the surface. A flexible rubber hose is attached to the spout of pump, which may be connected with the fuel receptacle of engine, permitting fuel to be pumped directly from the underground storage reservoir to the engine tank, without exposing it to the air. The pump may be kept under lock and key.

An efficient storage system permits the buying of fuel in large instead of small quantities, representing a considerable saving in the course of a year.

CHAPTER XVI

HOW TO SELECT A GAS ENGINE

The proper selection of a gasoline engine depends upon two general factors — the type of outfit that is best adapted to your particular requirements, and the make of engine that represents the greatest value for your money.

Below are outlined some of the principal points to be kept in mind and investigated in the selection of a gasoline engine.

The amount of power required for the work to be done. It is usually economy to get an engine a trifle larger than at first seems necessary, as more work is almost sure to develop than was originally contemplated. Not only this, but it is poor practice to have the engine so near the line of power required that it is often overloaded, as this is expensive both in wear and tear on the machine and in fuel consumed.

The nature of work to be performed is the factor which should determine the style of engine. If most satisfactory to have engine located permanently in one place and either bring machines to the engine, or install a line shaft for running the various machines, a stationary type of engine should be obtained. If the work is of such a character that it can best be handled by moving the engine to various points, one of the portable types, mounted on skids or wheels is best. If for hauling loads, and heavy duty work, threshing, plowing, etc., a tractor is best adapted.

A high grade engine is the best economy in the long run. Paint and polish, while in the strongest evidence to the untrained eye, are of very little consequence.

Ordinarily speaking, the gasoline engine with the fewest parts to do the work properly is the best, from the viewpoint of design, the same as with any piece of machinery.

The arrangement of parts so that they can be gotten at readily and easily removed is a good point.

As any piece of machinery will wear eventually, as well as suffer an occasional accident, it is of advantage to be able to get repair parts quickly, as delay in waiting for repairs keeps the engine out of commission and is often very expensive.

If engine is to be used indoors, investigation should be made to insure that it is well adapted for this service.

It is a point worth remembering that the rated horse power of an engine is not always a dependable basis for comparison. Many engines on the market rated at a certain horse power will barely pull this horse power at the best, while others with a generous rating, will if called upon stand an overload often as high as 15 to $33\frac{1}{3}$ per cent. above their rated power. If there is any doubt regarding the power possible to get from an engine under consideration, the prospective purchaser should demand proof of the actual brake horse power.

PART II

THE APPLICATION OF THE GASOLINE ENGINE

CHAPTER XVII

The general theory, principles and method of operating a gasoline engine are the same, whether engine is in an automobile, tractor, or truck, part of a water supply system or used as a stationary or portable outfit for general farm work.

The application of gasoline power, however, is varied and extensive. In this part of the book some of the most important applications for farming requirements will be taken up in detail.

CHAPTER XVIII

THE STATIONARY GAS ENGINE

By far the most universal application of gas power on the farm, at present, is the use of what is generally known as the stationary gas engine for running all kinds of light farm machinery.

Representative type of Stationary Engine.

Although called "stationary" engines, these outfits are generally more or less portable in nature, or at least provide for a certain amount of portability. However, there are outfits that are equipped and specially adapted for portable use, which will be taken up later on, and for the present we will consider the type of engine usually set on a stationary foundation.

Setting a Stationary Engine

The proper foundation is important to the successful operation of the engine. It should be solid and substantial so as to bear the weight of engine, perfectly even, and so constructed as to absorb vibration.

Concrete makes the most satisfactory foundation, although stone and brick are very satisfactory for the purpose.

The size and depth of foundation depends upon the size of engine.

As a general thing, if concrete is used, foundation should be from 3 to 5 ft. deep, a 3 ft. foundation being suitable for the smaller sizes from 5 horse power up, and the 5 ft. depth taking care of the largest sizes generally employed on the farm, from 25 horse power down. It is a good

Representative type of Stationary Engine.

rule to make the length of foundation twice the length of engine, at the bottom, and the width at bottom about 2¼ times the width of engine base. The foundation should be built up in pyramid shape so that it is about 6 inches larger than the engine bed at the top of foundation.

A foundation blue print should be obtained before starting to lay the foundation. The manufacturers of engine will generally furnish one without charge.

There is more or less variation in the details of construction, but the following general specifications of a 3 ft. foundation for a small engine may be taken as typical of a good, present day concrete foundation.

Obtain some long bolt rods with nuts and anchor plates —as many as there are bolt holes in bed of engine, and long enough to reach through engine bed down to within a few inches of bottom of foundation. These may have to be made specially by threading iron rod.

Some pieces of gas pipe should also be provided, a little larger in diameter than rod and long enough to reach from head of bolt up to, but not through, engine bed.

First, make a templet of boards (ordinary one-inch boards will do) the exact size top of foundation is desired to be. Then build a frame about 2 inches high and just a trifle larger than templet. Place frame and templet in place, taking care to square it just the way you want it to be and fasten it with stakes.

Then fill in the bottom of pit with concrete in about the proportion of 7 parts crushed stone or gravel to one of cement. Tamp it.

Fill in concrete, tamping it as you go along. The concrete will fill in around gas pipe and pack in on top of anchor plate, holding it firmly.

When concrete is filled in nearly to top unfasten nuts and remove templet. Fill up to the top with sand and cement, proportion about two to one.

Leave foundation to harden.

When thoroughly set, place the engine on foundation, working it along slowly and carefully with blocks, planks, rollers and levers, but keeping it above foundation high enough to clear protruding bolts.

When engine is approximately in place, lower to foundation slowly, so that bolts enter holes in engine frame.

After final lining up of engine, pour a fluid mixture of cement and water into the gas pipe sleeve around bolt, and let engine stand two or three days.

It should then be leveled up, to overcome any variation in the top of foundation and bottom of engine frame.

The crank shaft should be perfectly leveled by the use of a spirit level and by inserting thin wedges before the nuts are finally tightened.

CHAPTER XIX

POWER TRANSMISSION FROM GAS ENGINE TO DRIVEN MACHINE

After the engine is in place, the question of transmitting power to the machine to be operated is the next matter for attention.

There are various methods of installation, some of which are specially adapted to one purpose or outfit, which will be taken up later on.

There are, however, three general methods of connecting engine with the driven member which are adaptable and very extensivly used.

1. Belt Driven.
2. Direct Connected.
3. Gears.

Belt Drive

In the use of the gasoline engine on the farm, the transmission of power is almost altogether by means of belts running from a pulley on the engine either to a pulley on the machine to be driven or to a line shaft to which a number of machines are belted.

The individual belt drive from engine to single machine is the simplest method and most economical in first cost, in cases where various machines of one kind or another are operated one after another, and where it is possible to transport the machines to the engine (or the engine to the machine). Through the use of different size pulleys on the engine and driven member, the speed can be geared up or down to whatever point desired.

With belt driven outfits, the connection is easily made, and there is little to have attention either in the selection or use of transmission. The proper sizes of pulleys to operate any machine at its required speed can be ascertained from the manufacturers, or it can be figured from the tables in a later chapter on "Shafting, Pulleys, and Bearings."

The Line Shaft Drive offers the advantage of running several machines at once, instead of running one machine and doing one job at a time. By means of various sized pulleys, the various machines can each be run at its proper speed. Further information regarding the erection and use of line shafting will be found in another chapter on "Power House Construction," and data on pulleys and shafts is given in a following chapter on "Shafting, Pulleys, and Bearings."

General Installation Hints

It is desirable to locate engine so that belt will drive as nearly as possible to the center of shafting from which power is obtained. The center line of engine should be exactly at right angles to this shafting, and pulley of engine should be perfectly in line with the pulley which is to receive power from the engine.

Direct Connected

The direct connected drive can only be utilized to advantage for certain, particular kinds of work. It consists in connecting the engine direct to the machine to be operated, by some such method as a clutch or coupling which can be thrown on or off. In direct connected drives the engine and driven machine can obviously be operated only at the same revolutions per minute. The direct connected drive is only adapted for use where there is sufficient use for some power machine to hook machine and engine up permanently together, and keep the outfit for this one purpose only, and even then, as a rule, it would be more desirable to erect a line shaft and arrange for running other machines from the same power, except in cases where machine is so located that it would not be possible or desirable to locate other machines nearby or in the same building. For instance, in an independent water supply or electric light system it might be advisable to direct connect a small engine and keep it for this purpose only, also for irrigation. In direct connected outfits great care must be used to have the shafts of engine and machine in perfect alignment.

Gears

Gears, while used extensively on tractors, power mowers, trucks, automobiles, etc., will not be taken up here to any great extent for the reason that in the majority of instances where used, they are part of a complete outfit turned out by the manufacturers, and the farmer seldom has to work out any gearing problems.

In the operation of gear driven equipment, great care should be taken not to overload or throw a load on too suddenly, or the gears will be stripped of their teeth. There is absolutely nothing to give, whereas on a belt, overload will simply cause slippage and will do no permanent injury. Gears should always be kept free from dirt and well oiled, and worn gears replaced promptly before more serious injury is done.

CHAPTER XX

BELTING

It is important to satisfactory results that belting be not only of good quality but adapted to the purpose for which it is used.

Leather Belts

Leather belts stand rough handling and are well adapted for use on small pulleys. Oak tan leather is the strongest and most satisfactory for belts. Leather belts are unquestionably the best for all around farm use, but are very high priced. The hairy side being hardest and the fleshy side most flexible, the hairy side should be next the pulley.

Rubber Belts

Rubber belts cost less than leather and are uniform in width, thickness and strength. They withstand heat, cold and moisture better than leather and have the advantage of being made in any length with but one joint. They are less likely to slip than leather belts, but oil or grease is ruinous to them, and they will not stand up well under shifting. In strength a four-ply rubber belt is considered about the same as a single thickness of leather belt of the same width.

Canvas Belts

Canvas belts are uniform in thickness and strength, and cost less than either rubber or leather. They will stand moisture, oil, general hard usage, but become stiff in cold weather, contract with the weather, stretch and fray at the edges.

Use and Care of Belts

A belt should be slightly narrower than the face of a pulley.

A lapped belt should be put on so that the point of the lap runs off of the pulley and not onto it.

It should never be overloaded.

It should be kept clean and pliable.

Belt dressings should be used carefully and sparingly. Rosin should be avoided for leather belts; Neat's foot oil or beef tallow are best, in small quantities.

In cold weather it is a good plan to run for a few minutes without load to give belt a chance to get warmed up.

A belt should not be run too tight or too loose. An overtight belt binds and causes loss of power by friction, besides being expensive on belting and machinery. A loose belt slips and means loss of power. While excess slippage should be avoided a small percentage of slippage is a necessity, even an advantage, as sudden strains and overloads will simply cause the belt to slip, without serious after effects which usually result from such use in cog gear outfits.

Pulleys of about equal size will hold a belt with less slippage than where one very large and one very small pulley is used.

Horse Power Belting Will Transmit

Width of Belt, Inches	H. P. Per 100 Feet Belt-Velocity		Width of Belt, Inches	H. P. Per 100 Feet Belt-Velocity	
	Single Belt	Double Belt		Single Belt	Double Belt
1	.09	.18	12	1.09	2.18
2	.18	.36	14	1.27	2.55
3	.27	.55	16	1.45	2.91
4	.36	.73	18	1.64	3.27
5	.45	.91	20	1.82	3.64
6	.55	1.09	22	2.	4.
7	.64	1.27	24	2.18	4.36
8	.73	1.46	28	2.55	5.09
9	.82	1.64	32	2.91	5.82
10	.91	1.82	36	3.27	6.55
11	1.	2.	40	3.64	7.27

In the above table the belt is assumed to run about horizontally; the arc of contact of smaller pulley has been considered as approximately 180°. Any reduction of this contact will make proportional reduction of horse power. Both pulleys are also assumed to be approximately equal diameters.

Above table is based on Single Leather or 4-Ply Canvas Belt or Double Leather or 8-Ply Canvas Belt. A 5-Ply Belt need be only three-fourths as wide as a 4-Ply; and a 6-Ply only six-tenths as wide as a 4-Ply.

Length of Belt Required

To find the length of a belt required between any two

pulleys, add the diameters of the two pulleys in inches, divide by 2, multiply by 3¼. To this quotient add twice the distance between the centers of shafts, which will give the required length.

Speed of Belts

The speed of belts depends upon various considerations, such as required speed for operating machine, size of pulleys on hand, etc., and the only limitations that need be set is to avoid the extremes. The power transmitted by belting increases up to about 6,000 ft. per minute, after which it begins to diminish. However, it is far safer and more economical to operate at 4,000 to 4,500 ft., and as little as 600 ft. can be used with satisfaction.

CHAPTER XXI

SHAFTING, PULLEYS, AND BEARINGS

COPYRIGHT 1913
RUMELY PRODUCTS CO.

Wherever shafting is used it should be of good quality, and as such a small amount is ever used on a farm, there is no excuse for buying a poor grade.

The advantage of high quality in shafting is greater than the usual quality factors of durability and long life. It means a great reduction in the size of shaft that can be used and in the amount of power required to turn it. For instance, if a high grade steel shaft is operated at a certain speed, a cheap iron shaft to have equal strength would weigh several times as much and would require several times as much power to turn it.

If only one machine were driven from the line shaft, or all the machines were run at the same speed, it would be a simple matter to install the drive, but where a number of machines are driven with great variation of speed from extreme high to extreme low it presents a rather complex problem to determine the proper diameters and speeds for pulleys and shafts.

Wherever possible it would be a good plan to put the problem up to some one with engineering knowledge. Many of the machinery supply houses maintain a staff of engineers who will furnish information without charge to their customers. This, however, is not necessary. The following tables will enable any one with ordinary mechanical knowledge and patience to figure out a line shaft installation to meet their particular requirements.

Horse Power of Shafts for Given Diameter and Speed

This table is used in general practice for the transmission of power where shafts are properly supported.

When shafts are used for conveying power from one point to another, without bending strains of pulleys, gears, etc., the next smaller size may be used.

Diameter of Shaft, Inches	REVOLUTIONS PER MINUTE									
	100	125	150	175	200	225	250	300	350	400
1 3/16	2.4	3.	3.6	4.2	4.8	5.4	6.	7.2	8.4	9.6
1 7/16	4.3	5.4	6.5	7.6	8.6	9.8	10.8	13.	15.2	17.2
1 11/16	6.5	8.	9.7	11.2	13.	14.6	16.	19.4	22.4	26.
1 15/16	10.	12.5	15.	17.5	20.	22.5	25.	30.	35.	40.
2 3/16	14.	17.8	21.	24.5	28.	31.5	35.6	42.	49.	56.

With reference to above table, there are frequently special cases in which the engineer or designer must depart from set rules and use his judgment in determining the size of the shaft as well as the number and location of bearings.

Rules for Determining Diameters and Speeds of Pulleys and Shafts

To Find Speed of Driven Pulley

To find the number of revolutions of driven shaft, when diameter of driving pulley and its speed are known, multiply diameter of driving pulley by its number of revolutions per minute, and divide the product by the diameter of the driven pulley. The quotient will be the speed of driven pulley in revolutions per minute.

Example: Driving pulley is 10 inches in diameter, and makes 325 revolutions per minute. At what rate would it drive a pulley 5 inches in diameter?

$$\frac{10 \times 325}{5} = 650 \text{ revolutions per minute.}$$

To Find Diameter of Driven Pulley

To find diameter of driven pulley when diameter and number of revolutions per minute of driving pulley are known, multiply diameter of driving pulley by the number of its revolutions, and divide the product by the number of revolutions the driven pulley is to make.

Example: What would be the diameter of a driven pulley making 650 revolutions per minute if driving pulley is 10 inches in diameter, and makes 325 revolutions per minute?

$$\frac{10 \times 325}{650} = 5 \text{ inches in diameter.}$$

To Find Speed of Driving Pulley

To find number of revolutions of driving pulley when its diameter and the diameter and speed of driven pulley are known, multiply the diameter of driven pulley by its revolutions and divide the product by diameter of the driving pulley; the quotient will be the speed of driving pulley in revolutions per minute.

Example: The driven pulley has a diameter of 5 inches and its speed is 650 revolutions. How many revolutions must a driving pulley 10 inches in diameter make in order to give driven pulley the correct speed?

$$\frac{5 \times 650}{10} = 325 \text{ revolutions per minute.}$$

To Find the Diameter of Driving Pulley

To find the diameter of driving pulley needed to operate a driven pulley at any desired speed, multiply the diameter of driven pulley by the number of its revolutions per minute, and divide the product by the number of revolutions of driving shaft; the quotient will be the diameter of driving pulley required.

Example: A machine to be operated at 650 revolutions per minute, with driven pulley 5 inches in diameter. Speed of drive shaft is 325 revolutions per minute. How large a driving pulley should be used?

$$\frac{5 \times 650}{325} = 10 \text{ inches in diameter.}$$

Kinds of Pulleys

The majority of pulleys are constructed of iron, steel or wood. Iron pulleys are the most common in use, being

Wood Split Pulley.

more compact and neater in appearance than wood and cheaper than steel. Wood pulleys can be safely run at a considerably higher rate of speed than iron ones, and also hold the belt better. Wooden pulleys, especially in the larger sizes, are extensively made in split form, which permits them to be placed on shaft without slipping over end of shaft and disturbing other pulleys on the shaft. Steel pulleys are also made in split form, and are safer than where cast in one piece as well as more convenient to install.

Iron pulleys are cast in one piece and may be made with both straight and crown face, which latter is slightly higher in the middle. This is of assistance in keeping belt on pulley, as a belt will always hunt the high place in center, but is sometimes objected to because of throwing the whole load on the center of the belt. Crown faced pulleys can only be used when there is no belt shifting to do.

Securing Pulleys to Shaft

This is accomplished either by keying to the shafting or by a set screw. Keying is generally considered the best method, as if well done it will

Iron Pulley.

hold with the greatest security and safety. Care should be taken to cut key way in both pulley and shaft to match accurately, and to have key exact width of key way, so that key can be driven to a tight fit. Set screws are not generally so satisfactory, as they are prone to work loose, and after slipping a few times become threaded and never hold so well again. If allowed to project, set screws on a revolving shaft are also dangerous, as they are difficult to see and may injure hand painfully or catch in clothing while oiling a bearing.

Steel and wood-split pulleys clamp onto the shaft and do not require keys or set screws.

Tight and Loose Pulleys

The tight and loose pulley, as is generally known, consists of two pulleys of the same diameter, located side by side on a shaft, one of the pulleys being secured to shaft and the other one running loose on shaft. By shifting belt back and forth from loose pulley to tight one, the shaft and machinery may be started and stopped at will without interfering with the operation of engine or any other machinery on the same shaft.

Idle Pulleys

Idle pulleys are sometimes employed, and consist of a pulley installed between the engine and machine to be operated in such a way that it simply runs on the belt, its weight being sufficient to keep the belt tight. Its advantages are that the belt can be run comparatively loose, thereby reducing strain on both belt and bearings. Also, it is generally arranged so that it can be lifted off the belt leaving engine entirely free from the belt for starting purposes.

Friction Clutches

The friction clutch is often used for throwing power on and off a revolving shaft. Friction pulleys of various designs are obtainable which can be applied to any make of engine. Two representative types are shown in accompanying illustrations.

Type of Friction Clutch for throwing power on and off line shaft.

Bearings

A revolving shaft of any kind must be supported by a bearing. As a bearing, must very closely approximate a perfect circle, and as it necessarily wears out of true more or less rapidly with the revolving of shaft, some method of making bearings renewable must be provided. This is very extensively accomplished by babbitt metal bushings, which take the wear instead of the bearing itself, and are easily replaced when worn. In some cases, especially for light, high speed machinery, bearings instead of being babbitted are made of phosphor bronze or some similar high speed, anti-friction non-wearable metal, and in many cases ball bearings or roller bearings are used, which latter are the most efficient types of bearings but comparatively expensive. Split adjustable bearings are also sometimes used which have adjustable shims that can be removed to take up wear.

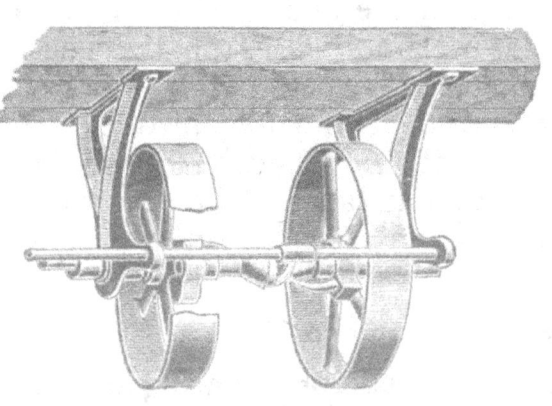

Another type of Friction Clutch, especially adapted for light line shaft work not to exceed 6 horse power.

Re-Babbitting

Bearings can be re-babbitted by those of fair mechanical ability if great care is taken in the work.

First remove old babbitt metal, and clean out all dirt and grease; then dry thoroughly.

Unless bearing is of the split type, wrap a piece of paper smoothly around shaft and gum the slightly overlapping edges to prevent babbitt metal sticking fast to shaft, allowing paper to project a little beyond edges of bearing.

Then place bearing in position on shaft, taking care to align perfectly, so that shaft is in perfect center of bearing, leaving a channel of uniform thickness around shaft, representing the babbitt bushing.

Close up ends of bearing with clay or putty, leaving air holes at the top, as well as a hole for pouring in the hot metal. Oil hole is sometimes used for this purpose, but if not a wooden plug should be inserted through oil hole in casting, with the end resting on shaft, to avoid the necessity of drilling it out afterward. A wall of clay or putty should be constructed around vent holes to keep them clear of flowing metal.

The babbitt metal, which can be secured at any machinery supply house, should be melted, heating it until it will brown or char wood. A regular plumber's furnace is very satisfactory for this purpose, as it can be carried about, permitting metal to be poured hot.

When heated to the proper degree, pour metal into the pouring hole until it begins to come up through air holes.

Care should be taken when pouring to see that no water or moisture which might generate steam is allowed to come in contact with the hot metal, as this is almost sure to blow out the babbitt metal and cause serious injury.

After filling, let it stand until cooled; then remove from the shaft, clean off clay or putty, trim off edges evenly, ream out oil hole with the end of a file if necessary, and cut a slanting groove across each side, for the oil to lie up against the shaft. This is important and must not be forgotten.

CHAPTER XXII

GRINDING FEED BY GASOLINE POWER

One of the most profitable things that can be done by gasoline power is the fitting up of a feed grinding room, instead of hauling to town, paying for the grinding and hauling back again.

It is agreed that stock will thrive better and put on more flesh in less time and with less feed when it is ground. It

Feed Mill belted to Gasoline Engine.

has also been shown by Agricultural Department and Experimental Station reports that there is a great deal of nutrition in ground cob and corn meal, thus utilizing for fodder what was formerly waste.

While the advantage of ground feed is recognized by all progressive farmers, the cost of having it ground, especially cobs, for which an additional charge is generally made by millers, and the inconvenience and loss of time in making trips to the mill in busy season, often prevent its use.

With the perfection of small feed mills, and the low cost of gasoline power, the farmer can grind his own feed at small cost. Not only this, but equipment is now obtainable which enables the farmer to make excellent wheat, rye and graham flour, corn meal, etc., representing a source of considerable additional saving by avoiding the necessity of buying these high priced food stuffs, as well as affording an opportunity to make money doing work of this kind for others in the neighborhood. Furthermore the same engine that produces power for operating a grinding outfit can be used for shelling corn, baling hay, cutting silage, sawing wood, etc.

The entire work of grinding can be made automatic by means of an elevator attachment operated by the engine power which conveys the grain, by means of an endless chain of conveyor buckets, from a portable bin set beside the thresher to elevated bins in the upper story of barn above stables. From these elevated bins, enclosed wooden chutes connect with the hopper of mill. By opening slides in chute the right distance, a stream of grain can be steadily fed to the mill just fast enough to be taken care of. The ground feed, as discharged from the mill, may be bagged or fed to a portable bin which can be moved along the floor and fed to feed chutes leading down to the feed boxes in the stable below.

The first consideration in obtaining efficient results from grinding equipment is in the selection of the mills. The buhr stones should be efficient in grinding qualities, permit of re-sharpening readily, and be easily adjustable for fine or coarse grinding.

The next essential is the proper application of power.

Although there is difference of opinion regarding the degree of fineness to which feed should be ground, the mill should grind with perfect uniformity and permit of close regulation, so that just the degree of fineness can be obtained that has been proven by experiment to be most nutritious and easily digested.

In the installation of grinding equipment, the size of mill

and horse power of engine depend upon the extensiveness
of the grinding. The capacity of the outfit, however, depends somewhat upon various conditions, such as whether
grain is hard or soft, dry or wet, the degree of fineness

Grinding feed from a Line Shaft.

desired, the ability of the operator, etc. There is no set
rule whereby the capacity of outfit can be determined with
exactness, but the following may be considered a fair average under ordinary conditions for feed, meal and flour.

MILL Diameter of Buhr Stones	GAS ENGINE Horse Power	BUSHELS PER HR.	
		Table Meal	Feed
14 in.	3 to 5 H. P.	5 to 8 Bu.	6 to 20 Bu.
18 in.	4 to 10 H. P.	8 to 12 Bu.	15 to 30 Bu.
22 in.	7 to 12 H. P.	10 to 18 Bu.	20 to 40 Bu.

If feed only is ground, of course the output will be more,
and if meal and flour only it will be less.

With reference to the application of power for operating
grinding equipment, it depends somewhat upon the location
of mill, which should be under cover. It can be so arranged that engine can be transported to the mill and belted
to it, or a line shaft or belt drive can be extended from
where the engine is used for other purposes into grinding
room or granary. Or the engine can be direct connected
to mill if there is a sufficient quantity of grinding to war-
rant the use of a small engine exclusively for this purpose.

The most satisfactory way, as a rule, to determine the

size of pulleys and other details of power application is to take the matter up with the manufacturers of mill, advising them how much power you have, what kind of power, the diameter of driving pulley and speed at which it runs, etc., and asking them to recommend the installation that will give best results with their particular machine. This, however, can be figured out for yourself by reference to chapters on " Shafting, Pulleys, and Bearings," " Belting," and " Power House Construction."

The use of a power feed grinder greatly facilitates the feeding of balanced rations, in which connection the following table of relative value of feed contents in one of the government agricultural bulletins will be found useful.

Feeding Stuff	Water Per Cent	Ash Per Cent	Protein Per Cent	Fiber Per Cent	Nitrogen Free Extract Per Cent	Fat Per Cent	Number of Analysis
Corn Silage	77.3	1.4	1.9	5.9	12.6	0.9	161
Red Clover	15.3	6.2	12.3	24.8	38.1	3.3	38
Alfalfa	8.4	7.4	14.3	25.0	42.7	2.2	21
Cowpea	10.7	7.5	16.6	20.1	42.2	2.9	8
Oat Straw	9.2	5.1	4.0	37.0	42.4	2.3	12
Oat Shorts	5.5	3.9	18.1	8.9	57.4	5.5	....
Kafir Corn	12.5	1.3	10.9	1.9	70.5	2.9	6
Barley	10.9	2.4	12.4	2.7	69.8	1.8	10
Oats	11.0	3.0	11.8	9.5	59.7	5.0	30
Oat Hulls	7.3	6.7	3.3	29.7	52.1	1.0	....
Rye	11.6	1.9	10.6	1.7	72.5	1.7	6
Wheat, Spring Varieties	10.4	1.9	12.5	1.8	71.9	2.2	13
Wheat, Winter Varieties	0.5	1.8	11.8	1.8	72.0	2.1	262
Wheat, All Varieties	10.5	1.8	11.9	1.8	71.9	2.1	310
Buckwheat	12.6	2.0	10.0	8.7	64.5	2.2	8
Buckwheat Hulls	13.2	2.2	4.6	43.5	35.3	1.1	....
Cotton Seed (with Hulls)	9.1	4.0	19.6	18.9	28.3	20.1	11
Cowpea	11.9	3.4	23.5	3.8	55.7	1.7	17
Corn Meal	15.0	1.4	9.2	1.9	68.7	3.8	77
Rye Flour	13.1	0.7	6.7	0.4	78.3	0.8	4
Ground corn & oats, equal parts	11.9	2.2	9.6	...	72.0	4.4	6
Corn Cob	10.7	1.4	2.4	30.1	54.9	0.5	18
Hominy Chop	11.1	2.5	9.8	3.8	64.5	8.3	12
Corn Bran	8.7	1.5	9.8	11.2	62.6	6.2	6
Corn Germ Meal	10.7	4.0	9.8	4.1	64.0	7.4	3
Oat Feed	7.7	3.7	16.0	6.1	59.4	7.1	4
Rye Bran	11.8	3.5	14.7	3.3	63.9	2.8	11
Wheat Bran, Spring Wheat	11.5	5.4	16.1	8.0	54.5	4.5	10
Wheat Bran, Winter Wheat	12.3	5.9	16.0	8.1	53.7	4.0	7
Wheat Bran, All Analysis	11.9	5.8	15.4	9.0	53.9	4.0	88
Wheat Middlings	12.1	3.3	15.6	4.6	60.4	4.0	32
Wheat Shorts	11.8	4.6	14.9	7.4	56.8	4.5	12
Wheat Screenings	11.6	2.9	12.5	4.9	65.1	3.0	10
Rice Hulls	8.2	13.2	3.6	35.7	38.6	0.7	3
Buckwheat Bran	11.5	4.5	24.8	11.7	40.8	6.7	7
Buckwheat Middlings	11.8	4.8	28.0	6.3	41.9	7.2	12
Cotton Seed Meal	8.2	7.2	42.3	5.6	23.6	13.1	35
Linseed Meal, New Process	9.9	5.6	35.9	8.8	36.8	3.0	33
Peanut Meal	10.7	4.9	47.6	5.1	23.7	8.0	2480
Peanut Hulls	9.0	3.4	6.6	64.3	15.1	1.6	5
Flaxseed	9.2	4.3	22.6	7.1	23.2	33.7	....
Peas	14.3	2.5	22.4	9.2	49.1	2.5	....
Dried Blood	8.5	4.7	84.4	...	...	2.5	....
Dried Fish	10.8	29.2	48.4	...	...	11.16	....

CHAPTER XXIII

THE CONCRETE MIXER ON THE FARM

Concrete mixers on the farm are coming to be used quite extensively.

Lumber is becoming scarcer and more high priced every year. A substitute has been necessary, and is being met by the use of cement for many purposes.

It doesn't take a very big job of cement construction to get back in labor alone the cost of a small concrete mixer that can be hooked up to the gasoline engine.

Every little while there is some job — laying a concrete floor in the barn or stable, hog house, or chicken coop, building a new foundation or cellar under the house, making a new watering trough or milk house — where an outfit of this kind is found to be of great advantage.

A concrete mixer is not only a great time and labor saver, but gives a much better grade of concrete than the hand mixing method.

There is usually an opportunity to make money with a concrete making outfit, by doing outside jobs, building bridges, foundations, tanks, etc.

A gasoline engine can be belt connected to any of the standard makes of concrete mixers on the market.

There are various types of concrete machines from the smaller machines that will mix a half yard of concrete at a time, to the large continuous mixers which will mix a continuous stream for an indefinite period. Light portable mixers mounted on wheels which can easily be drawn by one horse are usually of ample size for ordinary farm use.

CHAPTER XXIV

SAWING WOOD WITH GASOLINE POWER

In many localities considerable money can be saved by sawing your own wood by gasoline power, and there is often an opportunity to make a good income with a sawing outfit.

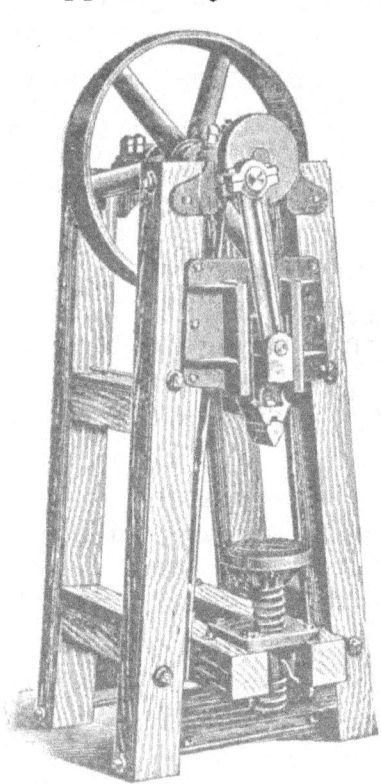

Chopping wood by power is now possible by means of an outfit like the above which is hooked up to a Gasoline Engine.

Outfits of every description can be obtained with wood or steel frames for sawing cord wood, stove wood, poles, railroad ties, fence posts, etc., with tilting or sliding tables, and drag or circular saws, from the small rig shown on page 104 to the elaborate saw mill on page 103.

Portable outfits are obtainable in which engine, saw and all equipment are installed in one complete mill, and others comprise the use of either a portable saw or a portable engine, or both may be portable and separate units.

The circular saw is in most extensive use for ordinary wood pile sawing, cutting up limbs, old rails, etc., and is most adaptable, although for long runs of large sized logs the drag saw can be used with less work.

Circular saw rigs can be purchased at very reasonable cost, or a good outfit may be put together in spare time by any one with ordinary mechanical ability. All that is necessary

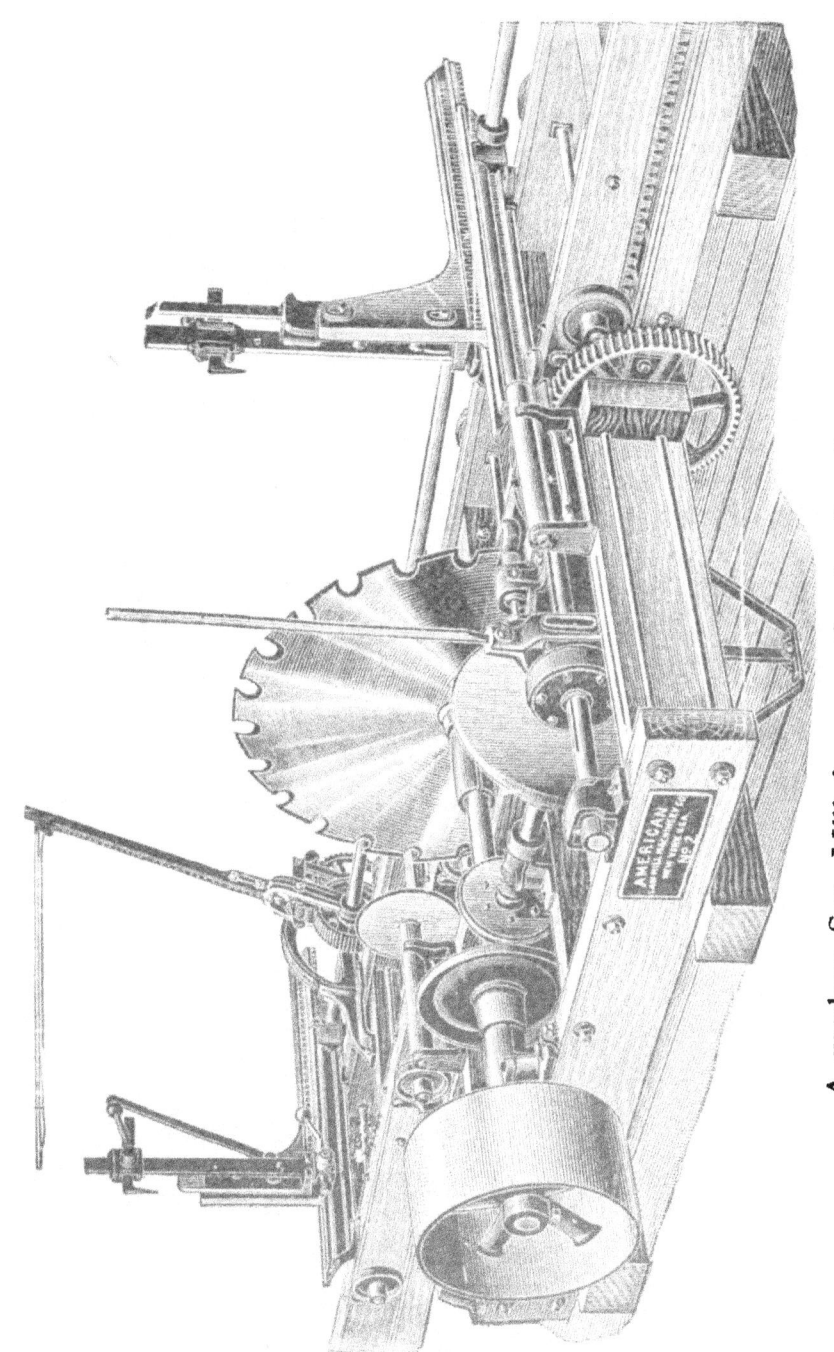

A modern Saw Mill for operation by Gasoline Power.

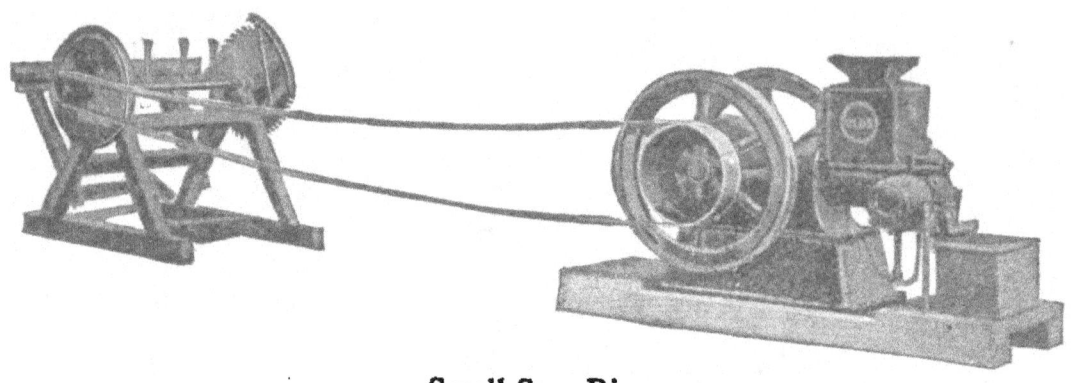

Small Saw Rig.

is a saw, saw guard and saw arbor, 4 or 5 feet of steel shaft, a wooden or iron frame mounted on a substantial foundation, a sliding or tilting table, a heavy balance wheel and a crown-face pulley.

A drag saw has the advantage that when once set in motion it will keep on going by itself, permitting the operator to split or pile wood or do other work.

CHAPTER XXV

HOISTING BY GASOLINE POWER

Various hoisting operations on the farm, such as unloading hay, hoisting bagged grain, etc., can be accomplished by a gasoline engine and hoist machine, made especially for the purpose.

Modern type of Power Hoist.

A typical machine of this character consists of two drums revolving on two steel shafts mounted on wooden skids. The rope drum has a capacity for several hundred feet of rope.

Unloading hay with a Power Hoist.

Hoist is driven by an iron chain from engine to sprocket wheel, engaging with gear on one of the drums. By pulling a lever, drums may be engaged with gears or disengaged at will. A brake is also provided.

CHAPTER XXVI

GASOLINE POWER, THE CHORE BOY

The gasoline engine has abolished a great deal of the drudgery of chore work and enabled the farmer to have more time for real productive work that makes the money on a farm, and for rest and recreation.

It has also made home more attractive to the boys on the farm, by eliminating this same irksome, monotonous chore work that is so distasteful to the boy with ambitious spirits, and giving him in its place the supervision of a piece of moving machinery that appeals to the natural mechanical instinct in most boys, and has a certain fascination for almost every person.

The gasoline engine has well been termed the farmer's right hand man. It does more work than two hired men can do in the same length of time. It is always on the job, in any weather, hot or cold, rain or shine, never complains at working overtime, and is never sick. The gasoline engine has done a great deal in solving the farm labor problem.

Portable outfit mounted on skids — can be carried by two men or dragged over the ground.

Portable Outfits

For chore work, a small portable outfit, of which there are various designs and types, is especially adapted.

These outfits permit the power plant to be moved

A wheelbarrow type of portable outfit.

around from place to place at a moment's notice, single handed.

For instance, outfit can be run out to the granary to grind the feed, shell corn, etc., wheeled over to the dairy to run the separator or churn, out to the wood pile to saw wood, or turn the grindstone.

The simplest form of portable outfit consists of wooden skids on which engine is set and bolted down, the ends of skids being fashioned into handles, permitting it to be picked up and carried about by two men, or on long hauls to hitch a horse to it and drag it over the ground.

Another form consists of wooden skids, with wheels mounted on one end like a wheelbarrow and the other end shaped into handles, which can be pushed along the ground like a barrow.

In another form, the engine is mounted on a four wheel steel or wooden truck with a handle in the front for hauling purposes.

It is not a bad plan and usually saves money in the end on a good sized farm to have two gasoline engines, one small portable outfit and a con-

Portable outfit mounted on Four-wheeled Truck.

siderably larger engine for doing heavier work, such as sawing wood, running the large grinder, sheller, shredder, silo, etc. The saving in fuel in operating a small engine instead of using a larger one for the light jobs will in many cases pay for a little portable outfit within a reasonable length of time.

Grinding Bone Meal for Poultry

To get best results from raising hens, they should be fed ground bone meal. Fresh cut green bones are very much relished by hens, and increase egg production as well as maintain the general vitality of the flock.

A bone grinder can be obtained at small cost that can be connected to any gasoline engine, and the bones ground every day in a few minutes, instead of burying or burning them.

Shelling Corn

With a gasoline outfit, you can shell your own corn, as much as you want at a time, and whenever you want to, without waiting for the contract sheller and paying a big price to get the work done, or doing the work by hand, a long drawn-out job that is usually more expensive in the end. Where it formerly took something like an hour and a half to shell a bushel of corn by hand, a gasoline outfit will do it in about one minute.

Husking Corn

You can husk your own corn, clean and without shelling the corn, upwards of 200 bushels a day, with a little power husking outfit, and make enough money to pay for the outfit in a short time by doing this work for neighbors.

Running the Grindstone

You can run your engine up to the grindstone, belt it up, start it going; and it will run itself, more evenly and steadily than by foot or hand power, and with much less work.

When grinding by power, have the outfit geared to run only fast enough for water to follow the stone without flying off. Remember to keep plenty of water in the trough, and when not in use drain off the water and cover stone. Watch cranks to keep them tight on the shaft.

Running Milking Machine

Milking is now done to some extent by gasoline power. It not only milks a cow much quicker than can be done by hand, but one man can milk two or four cows at a time. Although not so practicable in the early stages, a degree of efficiency has been reached that is quite remarkable. There is no question about the sanitary factor of the milking machine, as the milk is at no stage exposed to dust or odors of the stable; and cows as a rule become accustomed to it and stand quietly as for hand milking.

The principle of the milking machine is something on this order: Two rubber tubes terminate in four adjustable caps which attach to the cow, the other ends of tubes connecting with a vacuum device on the floor by means of which a suction is indirectly created by the engine power, drawing the milk from the cow to a covered pail. Glass windows in the tubes enable the operator to see if milk is flowing and as soon as stream stops it should be shut off.

Gasoline Engine vs. Windmill for Pumping

While pumping water for live stock is very much of a farm chore if it has to be done by hand, the subject of PUMPING is such a broad one that it will be taken up in another chapter.

In this connection, however, a brief comparison of the gasoline engine vs. the windmill will not be out of place.

When the windmill balks on a calm day, or is put out of commission by a wind storm, lightning or ice or sleet, the farmer is out of water supply unless he stays away from important work to pump by hand. Practically every high wind storm leaves a trail of broken gearing, and dismantled wheels, fans, wings, towers, etc.

Windmill power is also very unsteady, part of the time running so fast it almost jerks the pump out of the well, and at other times so slow that, if leathers and valves are worn, it will hardly throw a stream at all.

With the windmill there is also more or less danger that children may climb up on it and fall off, windmill may blow over and injure some one, and there is always a certain amount of danger in climbing thirty or forty feet to oil and care for the mechanism.

A gasoline engine can be purchased for less than the price of a windmill, can be started up on the coldest days in winter or the hottest days in summer, with the knowledge that it will keep on the job, at just the speed you set it, just as long as you want it to run.

Furthermore, the engine can be used for running practically every light machine on the farm, as well as for pumping, if desired.

Miscellaneous

There are countless uses to which gasoline power can be adapted about the farm, either in the nature of chores or odd jobs of one kind or another, new uses being found every day.

Whitewash can be sprayed much quicker than it can be brushed and even more effectively, as it is sure to penetrate into all the cracks and crevices.

Horses can be clipped and sheep sheared with clippers operated by engine power instead of hand power.

Lawns can be sprinkled, wagons washed and similar work done by means of an outfit made with a rotary pump and suction hose, which will throw a large stream of water a considerable distance, from cistern, watering trough or other source of supply.   Such an outfit is very desirable for fire fighting purposes.

CHAPTER XXVII

GASOLINE POWER FOR THE HOUSEWIFE

Never has there been a machine on the farm which has lightened woman's work as much as the gasoline engine.

Instead of a life of drudgery for the farmer's wife and daughter, they can now have free moments for recreation and the better things of life, and not be too worn out to enjoy them.

Gasoline Engine belted to Governor Pulley which in turn is belted to Separator.

With gasoline power has come culture, because the farmer's family have had more time to devote to music, to social enjoyment and entertainment.

The gasoline engine as well as being the farmer's right hand man is the housewife's handmaiden. There are countless jobs about the house that can be done with power just as well or better than by human hand, in a fraction of the time, and with a great saving, if not the entire elimination, of back breaking labor.

The Dairy

The Cream Separator is ordinarily run twice every day, or 730 times in a year. The gasoline engine can be used to save time and hard work every time it is run.

It is generally conceded that four cows and a good separator will produce at least as much cream as five cows without a separator. The Kansas Board of Agriculture makes the statement that in 1909, 3,331,960 pounds of butter fat, valued at over one million dollars, were saved by the use of centrifugal separator over old methods. The benefit of a separator stands undisputed. It will, however, make even more profit when operated by a gasoline engine than by hand, as it is almost impossible to turn the separator as steadily by hand as with power — and it is a real job to turn it fast enough.

With gasoline power, a governor-pulley is generally used, which permits the separator to be started gradually instead of too violently, and prevents it from varying during operation.

The Churn has come into more extensive use again with the advent of the gasoline engine, to the profit of the farm household. The hard work can now be done by power with very little interference with other household duties. With a little observation and experiment, the work can also be done in a more scientific way which will insure more efficient and uniform results.

Butter Working Machines may also be operated by power to good advantage.

The various machines in the dairy equipment may be operated, if desired, by the regular farm engine, run up to or into the dairy room and belted to each separate machine or to a line shaft to which the various machines are connected. As the power required for operating dairy equipment is small, 1½ horse power usually being ample, it is often found much more convenient and approximately as profitable to have a small size engine kept permanently in the dairy.

The Laundry

The gasoline engine has turned " Blue Monday " into a day of pleasure, comparatively speaking at least.

By hooking the washing machine up to a gasoline engine the washing can be gotten out of the way in about half the time it takes to do the job by hand, as the engine will be running the washer while part of the clothes are being put through rinsing water or hung out to dry.

The wringer also can be operated by the engine, leaving both hands free to handle the clothes and keep them from bunching, as well as saving the hard work of turning the wringer, which turns even harder than the washer.

Ironing also is sometimes done by power, by means of a mangle similar to those used in large public laundries.

The requirements of a household laundry can be met by an engine of 1 horse power, and on account of the low cost of an engine of this size many farms are equipped with a separate engine for the laundry and household purposes.

The Vacuum Cleaner

One of the most disagreeable but most insistent duties of the housewife is house cleaning. The vacuum cleaner will draw the dirt by suction from the corners of the room, instead of simply stirring it up to settle back again on floors and furniture.

The power vacuum cleaner makes it unnecessary to remove rugs and carpets from the house to clean them, and frequent cleaning in this easy manner does away with the periodical housecleanings which are looked forward to with dread by every member of the family.

With a vacuum cleaner curtains, draperies and hangings, as well as both men's and women's clothing, can be dry cleaned in a most efficient manner.

The vacuum cleaner is operated by hooking up to a gasoline engine. In many cases cleaning apparatus and engine have been mounted on a light wagon or truck, making a portable outfit with which to travel from house to house doing this work at a good profit.

Little Things About the House

There are numerous little things about the house which can be done by the gasoline engine to conserve the time and strength of the women of the household.

The sewing machine can be run with it, not only saving a lot of hard work, but enabling the operator to give more individual attention to the work.

It can be used for grinding the coffee, turning the ice cream freezer and operating the sausage grinder, bread mixer, etc.

In large families, where boarders are kept or large numbers of people eat, dish washing machines may be obtained which can be operated by gasoline power.

It can also be used to run an electric fan for cooling the kitchen, a sick room or other parts of the house, ventilating the basement, etc.

A very small engine is sufficient for household use, 1 horse power being ample; and, of course, it is much more convenient to have a small engine especially for household use than to hook up the farm engine, if it is desired to make systematic use of power about the house. The engine can either be hooked up to each machine individually or a small line shaft erected in the basement or some out room where all or most of the machines can be located.

Power House on the farm in connection with the granary.
(Illustration by courtesy of The International Harvester Co.)

CHAPTER XXVIII

POWER HOUSE CONSTRUCTION

Running a farm for profit is no different in principle than running a factory or any business establishment.

Good business management and up-to-date methods must be employed and efficient machinery used — the expense of production must be cut down to the lowest notch, and the highest percentage of results must be accomplished — if the farm is made to pay the biggest profits possible.

A Power House is the ideal method of using power on the average farm. It affords a number of advantages which contribute to the most profitable results from power.

The idea of the farm power house is to fit up part of a barn or other building, or to erect a small structure especially, in which the various power machines are placed and operated from line shafting by the engine.

Among the advantages of a power house are the following. The source of power can be centrally located. A great deal of time and labor can be saved in moving engine or machines when work is to be done. Instead of running one machine and doing one job at a time, several machines can be operated at once, thereby saving both time and fuel.

The power house provides a shelter for the engine and machines, and enables them to be easily kept clean and free from dirt.

It provides a convenient place for making repairs and improvements on days when it is impossible to work outdoors. It also affords a good opportunity to fit up a regular work shop with lathe, drill press, buzz saw, planer, grindstone, etc., as these machines can be run at the same time as the necessary farm machinery without additional cost for power.

Almost invariably when power houses have been installed, much larger profits have been made, and the power house has soon paid for itself in the great saving of valuable time.

In fitting out a power house, many things are necessary to consider — the general lay-out of floor space and location of the various machines and engine — the best plan of putting up line shafting — the speed at which to run shaft — the proper size of pulleys, etc.

The general arrangement of floor space obviously depends to a large extent on the size and shape of room and the number of machines to be operated.

Generally speaking, it is well to keep all machines of a certain character together, and better still to partition off sections for each kind. An excellent arrangement is to locate power house near granary, extending a shaft into granary for running feed mills, grinders, crushers, etc., then dividing power house into sections, one for a general workshop, one for laundry or dairy; one for general farm machinery, etc. It is also a very good plan to have a small enclosure partitioned off for the engine, especially if the machines used are going to raise a great deal of dust.

Ordinarily it is better, if possible, to install engine near the center of line shaft, instead of at one end, and to locate the heaviest machines nearest the engine.

After settling upon the arrangement that seems best suited to the particular requirements and conditions, the next thing to have attention is the erection of shafting.

Usually the best place for the shafting is near the ceiling where it will be out of the way and prevent accidents. The use of drop hangers attached to the ceiling is the most satisfactory arrangement, and if the shaft is of considerable length the hangers should be close enough together to reduce vibration and avoid possibility of shaft springing. In no case should hangers be over seven or eight feet apart.

For ordinary work a shaft about 1¼ inches in diameter is generally most satisfactory. When installing shaft be sure to have it run in a straight line and perfectly level.

The speed of line shaft should then be decided upon. The various machines to be driven generally operate at widely different speeds. This is accomplished by the use of different sized pulleys on the shaft and machines. To arrive at the proper speed for a line shaft requires careful figuring. For instance, if you intend to operate a cream separator, a slow speed machine, and also a feed grinder or other high speed machine from the same line shaft it will be necessary to run the line shaft at a slow speed to accommodate the separator and then use a large pulley on the shaft to drive the grinder. (Separators run at such extremely slow speed, however, that it is generally more practical to use a speed reducing pulley or governor than to attempt to run main line shaft at the slow speed necessary to operate separator.)

By proper manipulation of pulleys almost any machine used on the average farm, from the slowest speed machine up to machines that operate at as high as 1,000 revolutions per minute, can be driven from the same shaft. If conditions will permit such an arrangement, however, it is better to have an extra shaft driven from the main shaft for operating the slow speed machines.

The line shaft should, if possible, be run at a speed that will make it unnecessary to use any pulley smaller than three inches in diameter, or larger than thirty-two inches in diameter.

For rules covering the sizes of pulleys, proper belting to use, etc., see preceding chapters on " Shafting, Pulleys, and Bearings " and " Belting."

On any but a very short shaft, tight and loose pulleys for each machine are generally the most practical for farm use. This avoids the necessity of turning over the entire shaft and connections to start the engine, and permits the use of any one machine without starting the others. Friction clutches are also sometimes employed for this purpose, and as a matter of fact are as a rule more convenient and satisfactory to use, and cost but little more than tight and loose pulleys, if the latter are properly made.

CHAPTER XXIX

PUMPING

To pump water or other liquid against a head or resistance, power is required. It may be hand power, windmill power, gasoline power, steam power or other forms. Gasoline power is the ideal power for farm use.

The amount of power required per unit of time to move the given quantity of liquid from the source of supply to the point of delivery is called theoretical horse power.

Gasoline Engine belted to Pump Jack.

From this amount, an allowance must be made to cover power required to overcome the fluid friction in the pipes, and to overcome the mechanical friction of the pumping equipment.

The quantity which must be added to the theoretical power to cover this lossage depends upon various things, such as whether there are many sharp angular curves in the pipe line or whether it is perfectly straight, as well as upon the construction of the pump itself, etc.

The installation and operation of pumping equipment is based upon certain established rules and directions, as well as upon certain factors in connection with the particular kind of equipment used.

Following are given such rules and data as are generally applicable to power pumping:

General Directions for Installing and Operating Pumps

Pump should always be located as close to source of supply as possible.

Piping should always be run in as direct a line as possible.

Use as few elbows, tees and valves as possible.

Lay suction pipe with a uniform grade from the pump to source of supply, taking care to avoid pockets.

Centrifugal type of Pump, with primer,
equipped for belt drive.

Be sure that all connections and joints are absolutely air tight, as the smallest leak will supply the pump with enough air to prevent it working properly.

If source of supply contains foreign substances, which might get into and clog pipe, place a strainer on mouth of suction pipe.

All pipes leading to and from pump should be well supported to prevent undue strain on pump flanges.

Stuffing boxes should always be kept evenly packed, but never screwed down too tightly, as this causes an excess of friction and consequent loss of power.

In case pump does not start readily it may be due to air in the suction pipe and pump, in which case it may be necessary to prime pump cylinders with water and allow the air to escape through plugs or cocks provided for the purpose.

Pump valves should be examined occasionally to see if they are seating properly. Foreign substances are sometimes drawn in through suction pipe and lodge between valve and seat, causing leakage and reducing the force of pump. Metal valves should be ground occasionally.

All bearings should be kept well oiled, and should be examined frequently to see that they do not become loose or worn. A loose bearing may cause an endless amount of trouble and expense, and should be taken care of at once.

If pump is located where it is liable to freeze, it should be drained over night or when idle for a considerable period.

Useful Information

A U. S. gallon of water weighs 8.33 pounds, and contains 231 cubic inches or .133 cubic feet.

A cubic foot of water weighs 62.36 pounds and contains 7.48 gallons U. S., or 1,728 cubic inches.

A pound contains 27.7 cubic inches.

A cubic inch weighs .0361 pounds.

To convert Imperial gallons to U. S. gallons, multiply by the factor 1.2. To convert U. S. gallons to Imperial gallons, multiply by the factor .8333.

A miner's inch is a term used in measuring flow of water, and represents the quantity that will flow in one minute through an opening one inch square in a plank two inches thick under a head of 6½ inches to the center of orifice. It is approximately equal to 9 U. S. gallons per minute.

Atmospheric pressure at sea level is usually estimated at 14.7 pounds per square inch., and this pressure will maintain a column of water 33.9 feet high, if the vacuum is perfect. In practice, however, pumps should not be located over 25 feet above source of supply, and nearer if possible.

Every foot in height represents .434 pounds pressure to the square inch. In estimating, however, it is generally a good plan to figure one-half pound pressure per square inch for every foot in height, which will ordinarily allow for friction in pipes.

One pound pressure equals 2.31 feet in height.

To determine the theoretical horse power required to elevate water to a certain height, multiply weight of the water elevated per minute by the height in feet, and divide the product by 33,000 (allowance should also be made for water friction). Another rule extensively used is to multiply the gallons per minute by the head in feet and divide by 4,000.

The areas of circles are to each other as the square of their diameters. Doubling the diameter increases the capacity of pipe four times.

Friction of fluid in pipes increases as the length and velocity of flow increases.

In wooden pipes the friction is 1.75 times greater than in metallic.

To ascertain the velocity in feet per minute required to discharge a given volume of water in a given time, multiply the number of cubic feet of water by 144 and divide the product by the area of pipe in inches.

To find theoretical velocity due to any head multiply the square root of the head in feet by 8.02.

To determine the area of a required pipe, when volume and velocity of water are known, multiply the number of cubic feet of water by 144, and divide the product by the velocity in feet per minute.

STANDARD DIMENSIONS OF WROUGHT IRON PIPE AND LIGHT WELL CASING

Light Wrought Iron Well Casing				Wrought Iron, Gas and Water Pipe Standard Dimensions				
Inside Diameter	Outside Diameter	Threads per inch	Outside diam. of Coupling	Inside Diameter	Outside Diameter	Threads per inch	Diam. of Couplings	Length of Couplings
2 in.	2¼ in.	14	2¾ in.	⅛ in.	0.405 in.	27	41/64 in.	15/16 in.
2¼ "	2½ "	14	2 31/32 "	¼ "	0.54 "	18	23/32 "	1 "
2½ "	2¾ "	14	3¼ "	⅜ "	0.675 "	18	57/64 "	1¼ "
2¾ "	3 "	14	3 15/32 "	½ "	0.84 "	14	1 1/16 "	1 7/16 "
3 "	3¼ "	14	3¾ "	¾ "	1.05 "	14	1⅜ "	1⅝ "
3¼ "	3½ "	14	4 "	1 "	1.315 "	11½	1 43/64 "	1 15/16 "
3½ "	3¾ "	14	4 9/32 "	1¼ "	1.66 "	11½	2 1/32 "	2 "
3¾ "	4 "	14	4 17/32 "	1½ "	1.9 "	11½	2 11/32 "	2¼ "
4 "	4¼ "	14	4 25/32 "	2 "	2.375 "	11½	2⅞ "	2¾ "
4¼ "	4½ "	14	5 5/32 "	2½ "	2.875 "	8	3 9/32 "	3⅛ "
4½ "	4¾ "	14	5⅜ "	3 "	3.5 "	8	4 "	3 3/16 "
4¾ "	5 "	14	5⅝ "	3½ "	4. "	8	4 17/32 "	3⅜ "
5 "	5¼ "	14	5⅞ "	4 "	4.5 "	8	5 1/16 "	3¼ "
5 3/16 "	5½ "	14	6 1/16 "	4½ "	5. "	8	5 11/16 "	3⅝ "
5⅝ "	6 "	14	6 9/16 "	5 "	5.563 "	8	6¼ "	3⅝ "
6¼ "	6⅝ "	14	7¼ "	6 "	6.625 "	8	7 7/16 "	3½ "
6⅝ "	7 "	14	7 39/64 "	7 "	7.625 "	8	8 13/32 "	4 "
7¼ "	7⅝ "	14	8⅜ "	8 "	8.625 "	8	9⅜ "	4 "
7⅝ "	8 "	11½	8 23/32 "	9 "	9.688 "	8	10 29/32 "	6 1/16 "
8¼ "	8⅝ "	11½	9⅜ "	10 "	10.75 "	8	11 31/32 "	6 1/16 "
8⅝ "	9 "	11½	9 13/16 "	11 "	11.75 "	8		
9⅝ "	10 "	11½	11 "	12 "	12.75 "	8		

TABLE OF THEORETICAL HORSE-POWER REQUIRED TO RAISE WATER TO DIFFERENT HEIGHTS

Gal. per Min.	FEET HEAD																						
	5	10	15	20	25	30	35	40	45	50	60	75	90	100	125	150	175	200	250	300	350	400
5	.006	.012	.019	.025	.031	.037	.044	.05	.06	.06	.07	.09	.11	.12	.16	.19	.22	.25	.31	.37	.44	.50
10	.012	.025	.037	.050	.062	.075	.087	.10	.11	.12	.15	.19	.22	.25	.31	.37	.44	.50	.62	.75	.87	1.00
15	.019	.037	.056	.075	.094	.112	.131	.15	.17	.19	.22	.28	.34	.37	.47	.56	.66	.75	.94	1.12	1.31	1.50
20	.025	.050	.075	.100	.125	.150	.175	.20	.22	.25	.30	.37	.45	.50	.62	.75	.87	1.00	1.25	1.50	1.75	2.00
25	.031	.062	.093	.125	.156	.187	.219	.25	.28	.31	.37	.47	.56	.62	.78	.94	1.09	1.25	1.56	1.87	2.19	2.50
30	.037	.075	.112	.150	.187	.225	.262	.30	.34	.37	.45	.56	.67	.75	.94	1.12	1.31	1.50	1.87	2.25	2.62	3.00
35	.043	.087	.131	.175	.219	.262	.306	.35	.39	.44	.52	.66	.79	.87	1.08	1.31	1.53	1.75	2.19	2.62	3.06	3.50
40	.050	.100	.150	.200	.250	.300	.350	.40	.45	.50	.60	.75	.90	1.00	1.25	1.50	1.75	2.00	2.50	3.00	3.50	4.00
45	.056	.112	.168	.225	.281	.337	.394	.45	.51	.56	.67	.84	1.01	1.12	1.41	1.69	1.97	2.25	2.81	3.37	3.94	4.50
50	.062	.125	.187	.250	.312	.375	.437	.50	.56	.62	.75	.94	1.12	1.25	1.56	1.87	2.19	2.50	3.12	3.75	4.37	5.00
60	.075	.150	.225	.300	.375	.450	.525	.60	.67	.75	.90	1.12	1.35	1.50	1.87	2.25	2.62	3.00	3.75	4.50	5.25	6.00
75	.093	.187	.281	.375	.469	.562	.656	.75	.84	.94	1.12	1.40	1.69	1.87	2.34	2.81	3.28	3.75	4.69	5.62	6.56	7.50
90	.112	.225	.337	.450	.562	.675	.787	.90	1.01	1.12	1.35	1.68	2.02	2.25	2.81	3.37	3.94	4.50	5.62	6.75	7.87	9.00
100	.125	.250	.375	.500	.625	.750	.875	1.00	1.12	1.25	1.50	1.87	2.25	2.50	3.12	3.75	4.37	5.00	6.25	7.50	8.75	10.00
125	.156	.312	.469	.625	.781	.937	1.094	1.25	1.41	1.56	1.87	2.34	2.81	3.12	3.91	4.69	5.47	6.25	7.81	9.37	10.94	12.50
150	.187	.375	.562	.750	.937	1.125	1.312	1.50	1.69	1.87	2.25	2.81	3.37	3.75	4.69	5.62	6.56	7.50	9.37	11.25	13.12	15.00
175	.219	.437	.656	.875	1.093	1.312	1.531	1.75	1.97	2.19	2.62	3.28	3.94	4.37	5.47	6.56	7.66	8.75	10.94	13.12	15.31	17.50
200	.250	.500	.750	1.000	1.250	1.500	1.750	2.00	2.25	2.50	3.00	3.75	4.50	5.00	6.25	7.50	8.75	10.00	12.50	15.00	17.50	20.00
250	.312	.625	.937	1.250	1.562	1.875	2.187	2.50	2.81	3.12	3.75	4.69	5.62	6.25	7.81	9.37	10.94	12.50	15.72	18.75	21.87	25.00
300	.375	.750	1.125	1.500	1.875	2.250	2.625	3.00	3.37	3.75	4.50	5.62	6.75	7.50	9.37	11.25	13.12	15.00	18.75	22.50	26.25	30.00
350	.437	.875	1.312	1.750	2.187	2.625	3.062	3.50	3.94	4.37	5.25	6.56	7.87	8.75	10.94	13.12	15.31	17.50	21.87	26.25	30.62	35.00
400	.500	1.000	1.500	2.000	2.500	3.000	3.500	4.00	4.50	5.00	6.00	7.50	9.00	10.00	12.50	15.00	17.50	20.00	25.00	30.00	35.00	40.00
500	.625	1.250	1.875	2.500	3.125	3.750	4.375	5.00	5.62	6.25	7.50	9.37	11.25	12.50	15.62	18.75	21.87	25.00	31.25	37.50	43.75	50.00

* Add 50% to secure actual Horse-Power required for average conditions. Table by courtesy of Fairbanks, Morse & Co.

NOTE.—This table is based upon a mechanical efficiency of 100% for the pumping unit. To secure actual horse-power required for average conditions, add 50%. Where the pressure in pounds per square inch against which the pump has to operate, is known, instead of the head in feet, it can be converted into head in feet by multiplying by 2.31, as each pound per square inch of pressure is equal to the pressure of a head of water of 2.31 feet.

TABLE SHOWING EQUIVALENTS OF PRESSURE AND HEAD OF WATER

FEET HEAD OF WATER AND EQUIVALENT PRESSURE

Feet Head	Pounds per Sq. In.	Feet Head	Pounds per Sq. In.	Feet Head	Pounds per Sq. In.
1	.43	60	25.99	200	86.62
2	.87	70	30.32	225	97.45
3	1.30	80	34.65	250	108.27
4	1.73	90	38.98	275	119.10
5	2.17	100	43.31	300	129.93
6	2.60	110	47.64	325	140.75
7	3.03	120	51.97	350	151.58
8	3.40	130	56.30	400	173.24
9	3.90	140	60.63	500	216.55
10	4.33	150	64.96	600	259.85
20	8.66	160	69.29	700	303.16
30	12.99	170	73.63	800	346.47
40	17.32	180	77.96	900	389.78
50	21.65	190	82.29	1,000	433.09

PRESSURE AND EQUIVALENT FEET HEAD OF WATER

Pounds per Sq. In.	Feet Head	Pounds per Sq. In.	Feet Head	Pounds per Sq. In.	Feet Head
1	2.31	40	92.36	170	392.52
2	4.62	50	115.45	180	415.61
3	6.93	60	138.54	190	438.90
4	9.24	70	161.63	200	461.78
5	11.54	80	184.72	225	519.51
6	13.85	90	207.81	250	577.24
7	16.16	100	230.90	275	643.03
8	18.47	110	253.98	300	692.69
9	20.78	120	277.07	325	750.41
10	23.09	125	288.62	350	808.13
15	34.63	130	300.16	375	865.89
20	46.18	140	323.25	400	922.58
25	57.72	150	346.34	500	1154.48
30	69.27	160	369.43	1,000	2308.

COMPARATIVE MEASURES AND WEIGHTS

MEASURE AND WEIGHT EQUIVALENTS OF ITEMS IN FIRST COLUMN

Measures and Weights for Comparison	U. S. Gallon	Imperial Gallon	Cubic Inch	Cubic Foot	Pound
U. S. Gallon	1.	.833	231.	.1337	8.33
Imperial Gallon........	1.20	1.	277.27	.1604	10.
Cubic Inch	.0043	.00358	1.	.00057	.0358
Cubic Foot	7.48	6.235	1728.	1.	62.355
Pound	.12	.1	27.72	.016	1.

DISCHARGE OF WATER

GIVEN IN CUBIC FEET PER MINUTE, THE AREA OF THE STREAM BEING ONE SQUARE INCH

Head	Discharge	Head	Discharge	Head	Discharge	Head	Discharge	Head	Discharge
1	3.34	31	18.60	61	26.08	91	31.86	121	36.73
2	4.73	32	18.90	62	26.29	92	32.04	122	36.88
3	5.79	33	19.20	63	26.49	93	32.20	123	37.03
4	6.68	34	19.49	64	26.72	94	32.38	124	37.18
5	7.47	35	19.77	65	26.92	95	32.55	125	37.33
6	8.18	36	20.05	66	27.13	96	32.72	126	37.48
7	8.84	37	20.33	67	27.33	97	32.89	127	37.63
8	9.45	38	20.60	68	27.54	98	33.06	128	37.78
9	10.02	39	20.87	69	27.74	99	33.23	129	37.93
10	10.51	40	21.13	70	27.94	100	33.40	130	38.07
11	11.08	41	21.38	71	28.14	101	33.57	131	38.22
12	11.57	42	21.64	72	28.34	102	33.73	132	38.37
13	12.05	43	21.90	73	28.53	103	33.90	133	38.51
14	12.50	44	22.15	74	28.73	104	34.06	134	38.66
15	12.94	45	22.40	75	28.93	105	34.22	135	38.80
16	13.37	46	22.65	76	29.11	106	34.39	136	38.95
17	13.78	47	22.89	77	29.30	107	34.55	137	39.09
18	14.18	48	23.14	78	29.49	108	34.71	138	39.23
19	14.57	49	23.38	79	29.68	109	34.87	139	39.37
20	14.95	50	23.61	80	29.87	110	35.03	140	39.51
21	15.31	51	23.85	81	30.06	111	35.19	141	39.65
22	15.67	52	24.08	82	30.24	112	35.35	142	39.79
23	16.02	53	24.31	83	30.42	113	35.50	143	39.93
24	16.37	54	24.54	84	30.61	114	35.66	144	40.07
25	16.71	55	24.76	85	30.79	115	35.82	145	40.21
26	17.04	56	24.99	86	30.97	116	35.97	146	40.35
27	17.36	57	25.21	87	31.15	117	36.12	147	40.49
28	17.68	58	25.43	88	31.33	118	36.28	148	40.63
29	17.99	59	25.65	89	31.50	119	36.43	149	40.77
30	18.30	60	25.87	90	31.68	120	36.58	150	40.90

Table by courtesy of The Goulds Manufacturing Co.

TABLE OF EQUIVALENTS

Gallons per Minute	Miner's Inches of 9 G. P. M.	Cubic Feet per Minute	Gallons per Hour	Bbls. per Minute 42 Gal. Barrel
10	1.11	1.3368	600	.24
12	1.33	1.6042	720	.29
15	1.66	2.0052	900	.36
18	2.	2.4063	1,080	.43
20	2.22	2.6733	1,200	.48
25	2.78	3.342	1,500	.59
27	3.	3.609	1,620	.64
30	3.33	4.001	1,800	.71
35	3.88	4.678	2,100	.83
36	4.	4.812	2,160	.86
40	4.44	5.348	2,400	.95
45	5.	6.015	2,700	1.07
50	5.55	6.684	3,000	1.19
60	6.66	8.021	3,600	1.43
70	7.77	9.357	4,200	1.66
75	8.33	10.026	4,500	1.78
80	8.88	10.694	4,800	1.90
90	10.	12.031	5,400	2.14
100	11.1	13.368	6,000	2.38
125	13.8	16.710	7,500	2.98
135	15.	18.046	8,100	3.21
150	16.6	20.052	9,000	3.57
175	19.4	23.394	10,500	4.16
180	20.	24.062	10,800	4.28
200	22.2	26.736	12,000	4.76
225	25.	30.079	13,500	5.35
250	27.8	33.421	15,000	5.95
270	30.	36.093	16,200	6.43
300	33.3	40.104	18,000	7.14
315	35.	42.109	18,900	7.5
360	40.	48.125	21,600	8.57
400	44.4	53.472	24,000	9.52
450	50.	60.158	27,000	10.7
500	55.5	66.842	30,000	11.9
540	60.	72.186	32,400	12.8
600	66.6	80.208	36,000	14.3
630	70.	84.218	37,800	15.
675	75.	90.234	40,500	16.
720	80.	96.25	43,200	17.1
800	88.8	106.94	48,000	19.05
900	100.	120.31	54,000	21.43
1,000	111.1	133.68	60,000	23.8
1,125	125.	150.39	67,500	26.78
1,200	133.3	160.42	72,000	28.57
1,350	150.	180.46	81,000	32.14
1,500	166.	200.52	90,000	35.71
1,575	175.	210.54	94,500	37.5
1,800	200.	240.62	108,000	42.85
2,000	222.	267.36	120,000	47.64
2,025	225.	270.70	121,500	48.21
2,250	250.	300.78	135,000	53.57
2,500	278.	334.21	150,000	59.52
2,700	300.	360.93	162,000	64.3
3,000	333.	401.04	180,000	71.43

Table by courtesy of The Goulds Manufacturing Co.

TABLE SHOWING CONTENTS IN GALLONS OF SQUARE TANKS AND CISTERNS

Dimensions of Bottom in Feet	*1	4	5	6	7	8	9	10	11	12
					*Depth in Feet and Contents in Gallons					
4 x 4	119.68	479.	598.	718.	838.	957.	1077.	1197.	1316.	1436.
5 x 5	187.00	748.	935.	1202.	1309.	1516.	1683.	1870.	2057.	2244.
6 x 6	269.28	1077.	1346.	1616.	1885.	2154.	2424.	2693.	2968.	3231.
7 x 7	366.52	1466.	1833.	2199.	2566.	2922.	3299.	3665.	4032.	4398.
8 x 8	478.72	1915.	2394.	2872.	3351.	3830.	4308.	4787.	5266.	5745.
9 x 9	605.88	2424.	3029.	3635.	4241.	4847.	5453.	6059.	6665.	7272.
10 x 10	748.00	2992.	3740.	4488.	5236.	5984.	6732.	7480.	8228.	8976.
11 x 11	905.08	3620.	4525.	5430.	6336.	7241.	8146.	9051.	9956.	10861.
12 x 12	1077.12	4308.	5386.	6463.	7540.	8617.	9694.	10771.	11848.	12925.

* To ascertain the contents of a square tank or cistern of depth not given, multiply the contents of tank one foot deep as in table by the required depth in feet.

TABLE SHOWING CONTENTS IN GALLONS OF ROUND TANKS AND CISTERNS

Diameter in Feet	*1	4	5	6	7	8	9	10	11	12
					*Depth in Feet and Contents in Gallons					
4	93.99	376.	470.	564.	658.	752.	846.	940.	1034.	1128.
5	146.87	588.	734.	881.	1028.	1175.	1322.	1469.	1616.	1763.
6	211.50	847.	1058.	1269.	1481.	1692.	1904.	2115.	2327.	2538.
7	287.86	1152.	1439.	1727.	2015.	2303.	2591.	2879.	3167.	3455.
8	375.98	1504.	1880.	2256.	2632.	3008.	3384.	3760.	4136.	4512.
9	475.85	1904.	2379.	2855.	3331.	3806.	4283.	4759.	5235.	5711.
10	587.47	2350.	2938.	3525.	4113.	4700.	5288.	5875.	6462.	7050.
11	710.84	2844.	3554.	4265.	4976.	5687.	6398.	7109.	7819.	8531.
12	845.97	3384.	4230.	5076.	5922.	6768.	7614.	8460.	9306.	10152.

* To ascertain contents of a round tank or cistern of the above diameters, and of depth not given, multiply the contents of tank one foot deep by the required depth in feet.

CHAPTER XXX

ISOLATED WATER SUPPLY SYSTEMS

Running water throughout the house can now be had in country and suburban homes as well as in the city.

To have plenty of hot and cold water in the kitchen, bath room and dairy at a turn of a faucet is a luxury that every farmer owes to himself and family if he can afford it without actual hardship.

As a matter of fact the cost of a modern independent water supply system is no higher over a period of years than city water, when special assessments for putting in and maintaining city water, taxes for fire protection, water taxes, etc., are considered.

The country resident, however, must provide this service for himself, whereas in the city it is taken care of for the resident, but in so doing the country dweller has a great advantage over the town man. The water supply from wells or springs is usually much purer and more palatable as well as colder.

With the several good water supply systems now on the market, there is no more reason why the country home should be without running water than that this long established custom should be denied the city resident. In fact, the country resident has more uses for his water supply than his city neighbor. He does not have the protection of city fire pressure, and besides has need for a great deal of water for live stock and irrigation of crops in addition to household requirements.

Various types of water supply systems are available. The determination of the particular system to use depends somewhat upon the individual conditions in each case, such as quantity of water to be used, the source of supply, etc., and somewhat upon the personal preference of the owner.

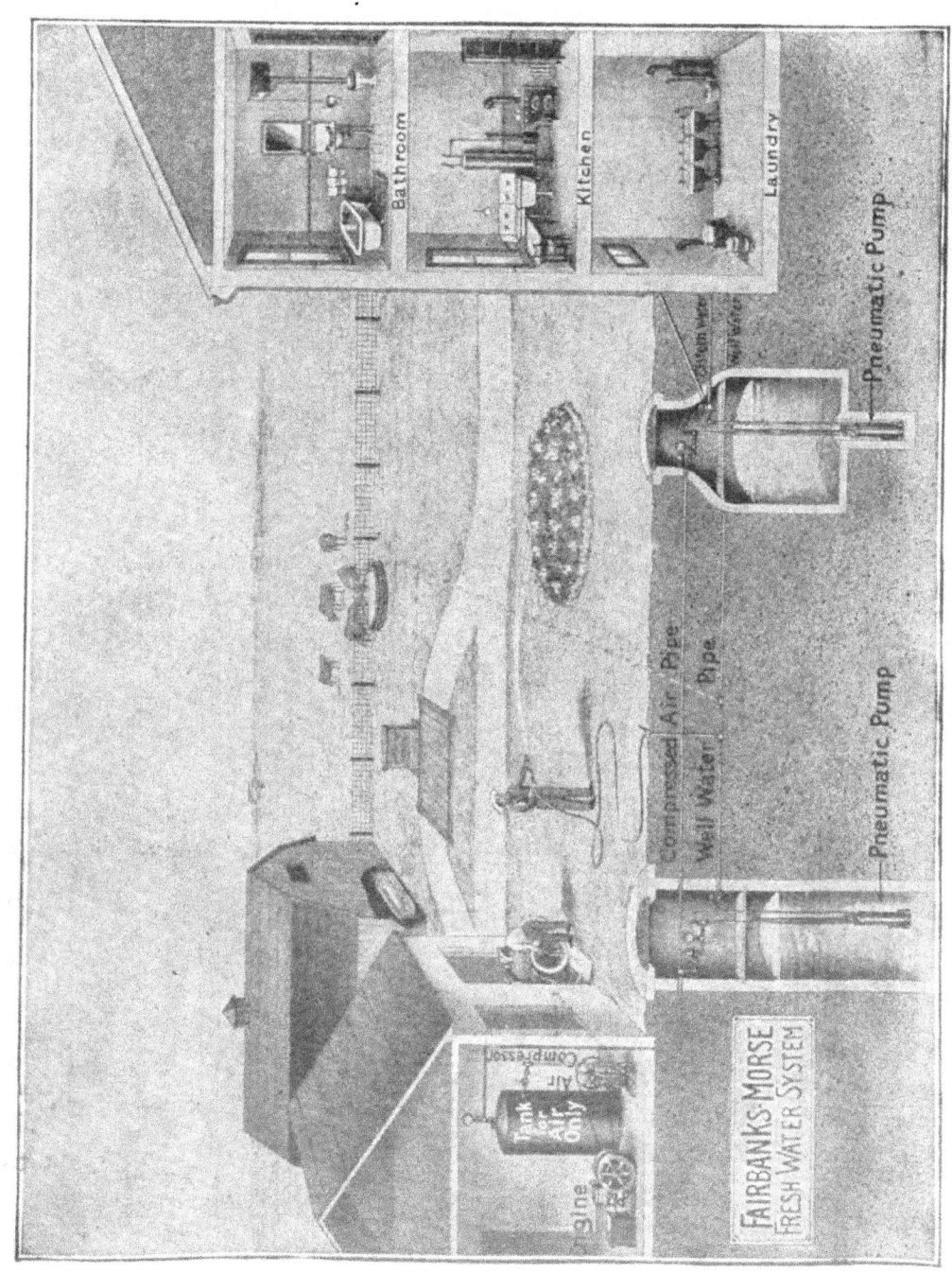

A representative "Pneumatic" Water Supply System. Illustration courtesy Fairbanks, Morse & Co.

Representative "Elevated Tank" Water Supply System. Illustration courtesy The Goulds Mfg. Co.

By far the majority of water systems get their supply from a shallow well, in which the water level is near enough to the surface so that water can be drawn up by suction, or from springs, lakes or rivers, which present similar conditions.

Deep wells are those in which the water level is too far below the surface to be drawn up by suction, requiring a plunger type of pump.

Two general methods are in quite extensive use for forc-

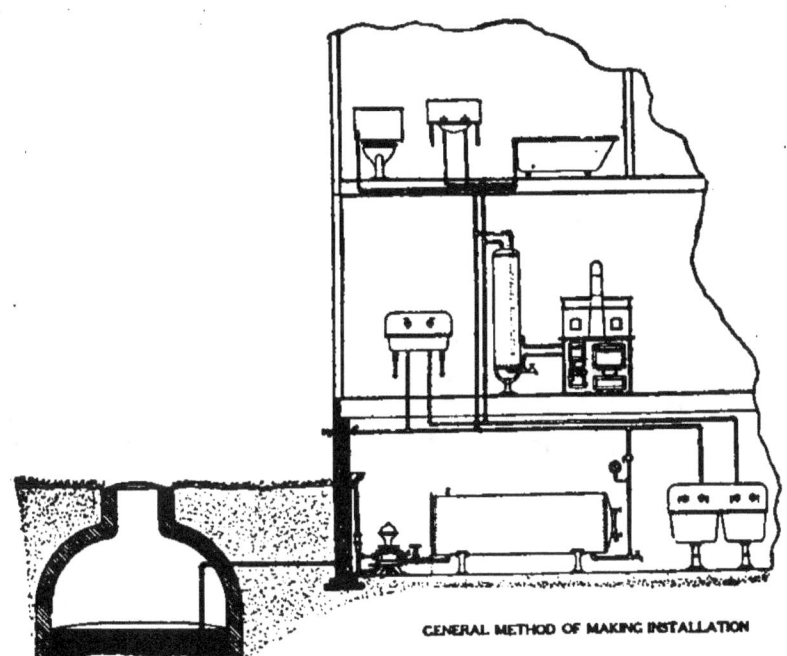

GENERAL METHOD OF MAKING INSTALLATION

General method of making installation " Pneumatic " or " Pressure " System.

ing water throughout the pumping system after it is pumped from the well or other source of supply.   In one, the water is forced into an elevated tank located higher than the highest faucet or discharge cock.   In the other, the water is pumped from the source into an air tight tank or receptacle usually located in the basement, and as a rule below the ground to keep it from freezing.   The water is forced up through the piping system by air pressure.   The air is sometimes furnished by a separate air pump and sometimes by the water pump.

The former system is commonly known as the " elevated

tank" system, and the latter as the "pneumatic," "pressure" or "compressed air" system.

Systems are also on the market whereby compressed air operates the pump, forcing water direct from pump through the piping system without storing it in a tank.

With any of the systems, provision may be made for heating water by some such method as running it through a water front in the fire box of kitchen range.

Points to be Considered in Providing Source of Supply

Water from springs, lakes and other surface sources, as well as shallow wells, should always be examined for impurities before being used for drinking. The best source of supply is a deep well, the water from a well 40 feet deep or over being pure and safe as well as cooler and generally more palatable.

A well should be located at least 100 to 125 feet away from any barn yard, cess pool or other source which might taint the water by seepage through the soil. It is also desirable wherever possible to situate it on higher ground than these impurities so that the natural flow of water will not be from them toward well.

The construction of a well has something to do with the purity and healthfulness of the water. It should be so built that surface water cannot get into it without having gone through a considerable layer of soil, filtering it and making it colder.

Kinds of Wells

There are three general types of wells in common use at present — dug, driven and drilled.

Dug Well

The dug well is the most primitive and the simplest type, in-as-much as it can be made by any one without necessity of special machinery. It is made by digging a large hole in the ground and lining it with masonry. Masonry should be made just as tight as possible for a depth of at least ten feet from the top, so that surface water cannot enter, and cementing it for this distance is very desirable construction.

The usual method is to extend masonry some eight or ten inches above the surface, a very satisfactory plan being to surround masonry with cement sloping from the top of masonry down to the ground. If a spout discharge is used and it is located directly over the water, a trough should be provided to carry the waste water away from the well.

Driven Well

Driven wells are more particularly adapted for shallow wells.

They are constructed by driving a pipe into the ground, on the lower end of which is attached a point. The lower section of piping is perforated with small holes through which the water can enter the pipe.

Driven wells can be used where the water is over 20 feet deep by digging a dry well down a certain distance, at the bottom of which to install the cylinders, and then driving pipe down from the bottom of dry well.

Drilled Well

This is the most satisfactory type of well. It is made by drilling a hole from 3 to 15 inches in diameter through the soil until good water is reached. The hole is lined with an iron casing. This keeps it from caving in and makes it impossible for any water to enter except at the bottom, absolutely keeping out surface impurities.

A drilled well can be sunk down through as many water streams as desired and only water from the lowest stream will enter the well.

A casing of sufficient size to accommodate pump cylinder is used, permitting cylinder to be located below the ground as near the water as necessary.

Useful Information

The water requirements for the various services ordinarily demanded of a water supply system are generally estimated about as follows:

Each member of family for all purposes,
 including kitchen, bath, etc..............25 gal. per day
Each horse ....10 " " "
Each cow ....10 " " "
Each hog .... 2 " " "
Each sheep .... 1 " " "

If the water is to be used in the house only, therefore, and there is a family of six, a water system should be used of ample capacity for $6 \times 25 = 150$ gallons per day. If water is to be used in both house and barn and there is a family of six persons, with eight horses, twelve cows, twenty hogs and ten sheep to be provided for, the water system should be ample for $6 \times 25 + 8 \times 10 + 12 \times 10 + 20 \times 2 + 10 \times 1 = 400$ gallons per day.

Where the tank system is used, if tank is located a considerable distance from the discharge cocks, tanks should be placed a little higher than otherwise, to make up for the loss of pressure in long pipe lines.

CHAPTER XXXI

IRRIGATION

Where nature fails to supply water in sufficient quantities it can be done efficiently and economically by gasoline power.

Irrigation has reclaimed thousands of acres of land in arid districts that formerly were absolutely worthless.

In regions subjected to drought, the irrigation of a small

Horizontal Centrifugal Pump direct connected to Gasoline Engine for irrigation work.

part of farm will insure a crop sufficient to carry the farmer over to next year.

In regions that normally have sufficient rainfall, countless instances can be pointed to where farmers, fruit growers and truck gardeners have been enabled to realize large increases in their yearly profit through the installation of an irrigation system.

In considering the installation of an irrigating plant, a number of points present themselves which must be correctly answered in relation to the individual conditions in each case. Among these points are the best method of applying the water, the proper quantity to use per acre, the frequency of application, the best type of pumping equipment, etc.

Gasoline Engine belted to Pump.

Necessarily the answer to these questions will vary greatly with the variable conditions of each section of the country, so that it is impossible to give any general recommendations that can be followed with certainty.

Before installing a system of irrigation, it is always advisable to make inquiries among local agricultural experts. The Departments of Agriculture of the various States have reliable data on irrigation which has been collected for the benefit of agricultural residents of their States. Every one who contemplates putting in an irrigation plant or who is now operating one, should write his State Department of Agriculture to send him all available information on this subject.

Water for irrigating purposes is provided by one of two methods: either by gravity flow from a source of supply located higher than the land to be irrigated; or by pumping it from a lower or practically even level.

Where it flows by gravitation, water is usually stored by means of a dam and carried to the irrigated land by flumes. Where the supply is not located higher than the irrigated land, or where the soil is too porous to make an artificial reservoir feasible, the water is pumped into the distributing flumes, ditches or pipes.

The type of pumping equipment to use and the method

of installation depend upon such conditions as whether water is from surface supply (lakes, rivers, shallow wells, etc.) or deep wells, upon the quantity of water required per hour, the height and distance water has to be pumped from the source of supply, and the character of water, whether clear or muddy and gritty.

Centrifugal pumps are in most general use for irrigation work, they being best adapted for lifting large quantities

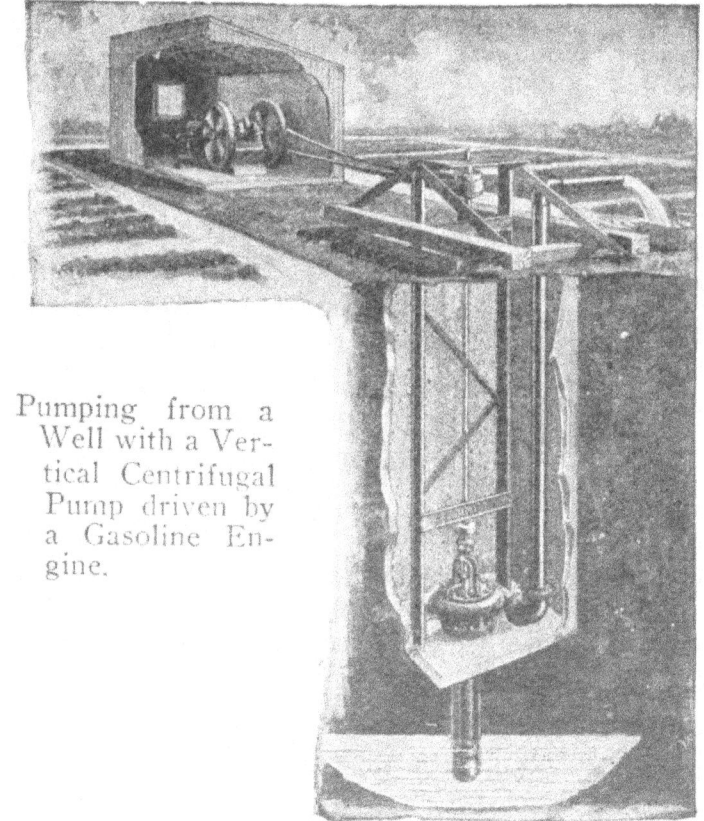

Pumping from a Well with a Vertical Centrifugal Pump driven by a Gasoline Engine.

of water a short distance, and less likely to clog with muddy water and foreign substances than valve pumps. Centrifugal pumps are of two general types — vertical and horizontal, accordingly as the shaft is vertical or horizontal.

Gasoline engines are economical and reliable drivers, and are extensively used for irrigation service in all sections.

Pumps may be either direct connected to engine or belt driven.

Care should be taken to have the engine of proper size

for the pump — if too small it will not run pump at a high enough rate of speed to be economical, and if too large it will waste power.

Under ordinary conditions a 10 to 15 horse power engine will operate a No. 4 or No. 6 centrifugal pump with a lift of 15 to 25 feet, and discharge about 2 cubic feet of water a second.

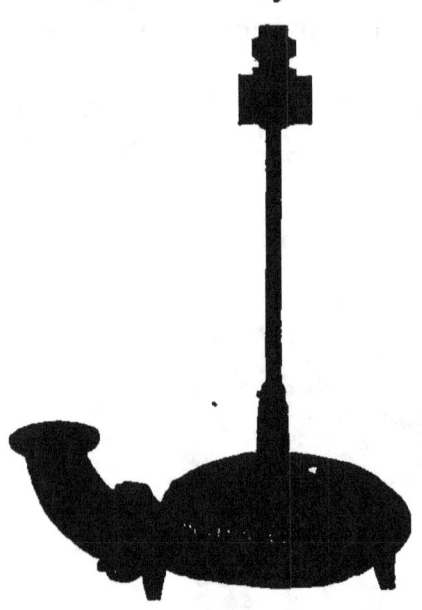

Type of vertical centrifugal pump.

The tables and data given in a preceding chapter on " Pumping," also in chapters on " Shafting, Pulleys, and Bearings," will be found useful in the consideration of irrigation problems.

After the water is pumped to the point of usage, it is distributed in various ways. In the great western irrigation plants, elaborate ditches and paraphernalia are provided. In some of the simpler plants, the water is pumped into a moderately elevated tank from which distribution is made, as the water direct from pump is likely to be cold, and by standing exposed to the sun for awhile the temperature becomes more suited for application to growing vegetation.

For garden crops and plants in rows, best results are generally obtained from root irrigation, which may be accomplished by running water into trenches on each side of the row of growing plants.

For grass and other close growing plants application from above is usually best, which may be accomplished by a garden hose and spray nozzle, or other apparatus.

To Measure the Flow of Water

The flow of water from a pump or supply stream can be measured by constructing a weir, which is in fact a rectangular notch cut in a dam across stream or in the side of a box into which the pump discharges.

The notch should be centered and its width should not be more than three-fourths the width of box. It should be wide enough so that water flowing over the lip will not be over 10 inches or less than 1 inch deep. The depth of water flowing over the weir must not be more than one half the drop outside. The edges of the notch should be beveled on the side and top. The box should be long enough so that water will be quiet before it reaches the

A common method of constructing a Weir.

weir and wide enough so that the flow will not be too rapid.

Four feet back from lip of weir a post is set with its top exactly level with the lip. The depth of water flowing over lip can be measured with an ordinary rule set on top of post.

GALLONS PER MINUTE FLOWING FOR EACH INCH LENGTH OF WEIR, WITH THE GIVEN DEPTHS AT THE LIP

Inches Depth	Gallons Per Min.	Inches Depth	Gallons Per Min.	Inches Depth	Gallons Per Min.
1	3.06	4⅛	25.47	7⅛	57.78
1⅛	3.60	4¼	26.64	7¼	59.31
1¼	4.22	4⅜	27.81	7⅜	60.84
1⅜	4.86	4½	28.98	7½	62.37
1½	5.58	4⅝	30.24	7⅝	63.99
1⅝	6.30	4¾	31.41	7¾	65.52
1¾	7.02	4⅞	32.67	7⅞	67.14
1⅞	7.83	5	33.93	8	68.76
2	8.64	5⅛	35.28	8⅛	70.30
2⅛	9.45	5¼	36.54	8¼	72.00
2¼	10.26	5⅜	37.89	8⅜	73.62
2½	11.97	5½	39.15	8½	75.24
2⅝	12.96	5⅝	40.50	8⅝	76.95
2¾	13.86	5¾	41.85	8¾	78.66
2⅞	14.85	5⅞	43.29	8⅞	80.28
3	15.75	6	44.64	9	81.99
3⅛	16.74	6⅛	46.08	9⅛	83.70
3¼	17.82	6¼	47.43	9¼	85.41
3⅜	18.81	6⅜	48.87	9⅜	87.21
3½	19.89	6½	50.31	9½	88.92
3⅝	20.97	6⅝	51.75	9⅝	90.72
3¾	22.05	6¾	53.28	9¾	92.43
3⅞	23.13	6⅞	54.72	9⅞	94.23
4	24.30	7	56.25	10	96.03

Frequently it is necessary to correct for rapid flow, or velocity of approach. This can be measured by timing sticks floating on the surface. The following table shows proper amounts to add to measured depths for different velocities:

Velocity of approach, in feet per second.	.5	1	1.5	2
Addition to measured depth	1–16	¼	⅝	1⅛ inch

The use of the above table is made clear by the following example:

Suppose the notch is 60 inches wide and the rule shows the depth to be 6 inches. Then the total flow will be 60 × 44.64 = 2,678.4 gallons per minute.

Table by courtesy of The Goulds Manufacturing Co.

IRRIGATION QUANTITY TABLES

	Amount of water required to cover one acre to given depths.		* Second Feet reduced to Gallons and Acre Feet.				Gallons required to cover a given number of acres to a depth of one foot (Acre foot)	
Depth in inches and feet (Acre inches and acre feet)	Cubic feet (or second feet) contained in one acre to depths given in first column	Gallons	Second feet	Gallons per minute	Gallons per pumping day of 12 hours	Acre feet per pumping day of 12 hours	Acres (or number of acre feet)	Gallons
ft. in.								
1	3630	27154	¼	112.2	80790	.2479	1	325851
2	7260	54309	½	224.4	161579	.4959	2	651703
3	10890	81463	¾	336.6	242369	.7438	3	977554
4	14520	108617	1	448.8	323158	.9917	4	1303406
5	18150	135771	1¼	561.0	403948	1.2397	5	1629257
6	21780	162926	1½	673.2	484738	1.4876	6	1955109
7	25410	190080	1¾	785.5	565527	1.7355	7	2280960
8	29040	217234	2	897.7	646317	1.9835	8	2606812
9	32670	244389	2½	1122.1	807896	2.4793	9	2932663
10	36300	271542	3	1346.5	969475	2.9752	10	3258515
11	39930	298697	4	1795.3	1292634	3.9669	15	4887772
1 00	43560	325851	5	2244.2	1615792	4.9586	20	6512029
1 2	50820	380160	6	2693.0	1938951	5.9503	25	8146286
1 4	58080	434469	7	3141.8	2262109	6.9421	30	9775544
1 6	65340	488777	8	3590.6	2585268	7.9338	40	13034058
1 8	72600	543086	9	4039.5	2908426	8.9255	60	19551087
1 10	79860	597394	10	4488.3	3231585	9.9173	80	26068116
2 00	87120	651703	20	8976.6	6463170	19.8345	160	52136232

* One cubic foot of water per second (exact 7.48052 gallons) constant flow is known as the " Second Foot." The " Acre Foot " is the quantity of water required to cover one acre to a depth of one foot.

CHAPTER XXXII

SPRAYING

Insect pests, it is estimated, do more harm to our forests than is done by forest fires, the loss from which runs into many millions of dollars annually.

Orchard Spraying.

If such extensive damage is done to sturdy forest trees, the amount of damage to less hardy orchard trees and to frail garden and field crops must be tremendous.

Spraying has already become the chief essential to successful fruit growing, and within the last few years has come to be recognized as an important factor in successful truck farming. It is also resorted to extensively in tobacco raising. Farmers growing grain and hay crops, however, have not yet taken up spraying so generally, although it has been shown by the experimental work of the Agricultural Department, State Experimental Stations and Agricultural Colleges, that spraying will just as surely increase the profit from these crops as from fruit.

The use of proper insecticides is a matter of more importance than the spraying equipment, as any of the well known standard outfits on the market can be depended upon to give generally satisfactory results, while, as a rule, failure in spraying is due to mixture not being properly made, the wrong formula being used, vegetation not sprayed at the right time or some similar cause.

The Federal Government, through the Department of Agriculture, disseminates knowledge on this subject, in the form of bulletins, that are sent free on application to any resident of the United States. More complete data

Spray Pump connected to Gasoline Engine by Gear.

regarding special conditions in any particular section of the country can also be obtained from the State Department of Agriculture. This service is free to residents of the State, and as it is supported by the taxes of the people, no hesitancy should be felt in asking for it.

Spraying Field Crops.

In addition to preventing crop failure and increasing the yield, spraying will enable the farmer to command a higher price for his products. Statistics kept by the Missouri State Board on the 1911 apple crop showed the average price per bushel from sprayed orchards was 61.7 cents, while the average price per bushel from the unsprayed orchards was but 49.3 cents. The yield was also shown to be increased an average of two to three bushels a tree. Data compiled by the Nebraska Experiment Station showed a net gain of $64.55 per acre in the value of sprayed crops over unsprayed. Statistics on the potato crop for a period of twenty years, kept by the Vermont Experiment Station, show an average gain of 105 bushels per acre or 64 per cent. in the sprayed crop over the unsprayed one.

Spraying is not necessarily a power operation, but if there is a large acreage to spray the work can be much more conveniently and economically done, as well as more efficiently by operating the spray pump with a gasoline engine. The spray thrown by hand operation is not so uniform or fine, and will not penetrate so thoroughly into heavy foliage or branches as power thrown spray.

Spray Pump Belt-connected to Vertical Gasoline Engine. Makes a very convenient outfit mounted on Skids as shown in illustration.

Various methods of spraying are employed, depending somewhat upon the crops.

For orchard work a very good method extensively used is to mount spray pump, engine and tank of insecticide on a wagon with a long hose attached, enabling the operator to have ready access to all parts of the tree. Often a platform is also provided, raised above the wagon six or eight feet for operator to stand on.

For potato and field crops, spraying attachments are provided which hang from the end of wagon down close to the vegetation, throwing spray directly on foliage.

Spraying should always be done with the greatest thoroughness. This does not necessarily mean that the foliage should be drenched with the solution; in fact with most mixtures this is bad practice, as it makes too poisonous an application. For best results, every particle of foliage should be reached with a fine mist. One method commonly employed in spraying trees is to throw the spray up into the air over the tree and let it settle down into the foliage. It is also important that all branches of a tree be reached, as a single unsprayed limb may undo the work done on the remainder of tree.

Another adjunct to efficient results is an agitator in the tank solution, which is some sort of mechanical contrivance kept moving about to prevent settling and keep the mixture of uniform consistency and strength.

For extensive and most scientific spraying an assortment of nozzles should be had, as various mixtures and conditions call for slight variations in the spray. For instance, lime-sulphur solution will stand a more thorough drenching and a coarser nozzle; and spraying after petals should be done with a little greater force than before.

Hose should be adapted to standing high pressure; $\frac{3}{8}$ or $\frac{1}{2}$ inch, 3 or 4 ply is generally used. Sometimes $\frac{1}{2}$ inch gas pipe or hollow bamboo is used, often being made in six or eight foot lengths that can be coupled together.

Other Uses for Spray Pump

A spray pump outfit will be found useful for a number of other services about the farm.

Barns and outbuildings can be painted with it or cellars whitewashed easily and quickly.

Greenhouse and hotbed glass can be quickly coated to produce any degree of opaqueness by using a fine or coarse spray.

Hen houses can be sprayed with a disinfecting liquid to rid them of vermin.

In fly time, stock can be sprayed with a mixture which will furnish protection from these pests.

Some Suggestions Regarding Spraying

Never spray foliage in excessively hot sunshine as it may cause leaves to fall or do other injury.

Spray when the leaves are dry. Poison is more effective when applied to dry leaves.

Do not spray directly after a shower or heavy dew, as much of the solution may be washed off.

Best results are obtained if spraying is commenced before pests have attacked vegetation; afterward it may be too late to save crop.

Spraying solutions and mixtures containing copper sulphate, corrosive sublimate, arsenate of lead, etc., should be made in wooden, glass or earthen vessels.

All substances used in making spraying mixtures should be carefully labeled and be kept in some place where they cannot be used by mistake.

Pump and fittings should be cleaned thoroughly after using by forcing clear water through them.

In cold weather water or liquid should not be left in pump, tank or engine over night. It might freeze and expand, doing serious damage.

CHAPTER XXXIII

ISOLATED ELECTRIC SYSTEMS

Another convenience of city life which the gasoline engine has helped to bring to the country and suburban residence is electric lighting.

The superiority of electric illumination is well known. Its advantages are almost countless.

Electric lights are convenient — a simple twist of a button turns on the light; no looking for matches or feeling around in the dark for a lamp or lantern. They are no bother to take care of — no filling or caring for lamps. They are safe — no dropped matches or knocked over lamps to start a fire or explosion. They are clean — no blackened chimneys, smoke or disagreeable odors

In cities electric lights, because of their convenience, cleanliness and safety, are in almost universal use. In the country, however, they have been practically unavailable before the advent of the gasoline engine, and even in the early days the cost of apparatus for generating electricity limited its use to those who were comparatively well-to-do. Within the last few years, however, improvements have been introduced which make it possible to obtain small electric light outfits at a cost within the reach of the great majority of farm dwellers.

The Lighting Outfit

With the most generally used systems, an electric lighting plant consists of a gasoline engine, dynamo and switchboard with storage battery reserve.

Engine

The engine is the source of mechanical energy for the operation of dynamo. Any standard make of gasoline en-

Typical installation of an Electric Lighting Plant on a Farm, showing the lighting of the Dwelling, Barn, Pump House and Tool House. The Electric Lighting Plant is set up in a corner of the Tool House. Illustration shown by courtesy of The Electric Storage Battery Co.

Complete Isolated Electric Lighting Plant.

Gas Engine and Dynamo of an Electric Lighting Plant.

gine adapted for electric lighting plants can be utilized for running farm machinery, pumping, etc., reducing the cost of lighting equipment very materially.

Dynamo

The dynamo is the source of electricity. When belted or direct connected to the engine it generates electrical current and delivers it at the switchboard.

Switchboard

The switchboard is a panel on which the apparatus for controlling and switching the current on and off is placed. From switchboard the current is carried over wires to the various electric lights, to the storage battery, if used, and if desired to motors which may be used for operating various machines.

Storage Battery

As the dynamo will deliver the electrical energy it generates only while in operation, a storage battery is usually employed which acts like a reservoir, storing the current and making it unnecessary to run engine and dynamo continuously in order to have electric light.

The frequency with which engine and dynamo must be run and the length of time required to charge a storage battery depends upon the size of the storage battery and the amount of current needed for the operation of lamps or motors after the engine is shut down.

Storage batteries used for small lighting plants ordinarily consist of a number of glass jars containing the electrolyte solution in which lead battery plates are placed, composing a cell. Cells are usually placed on shelves of two tiers.

Location of Lighting Plant

Complete lighting plants are generally set up in a corner of basement or out building. The plant should be located in a clean, dry place affording good light and ventilation.

Wiring

Various methods of wiring a house for electric lights may be employed, and the cost of the work depends upon how it is done.

The best method is to run the wires between partitions, with outlets in the walls or ceilings at desired places. This makes all wiring out of sight, but means considerable trouble except in cases of new buildings when it can be provided for when building.

Another method is to run wires in wooden moldings which are finished to match the woodwork, and are not conspicuous.

In places where the appearance of wires is not objectionable, such as in attics, cellars, barns, stables, garages, etc., exposed wiring is generally used.

Wiring, where approximately ten or more lights are installed usually costs in the neighborhood of $2.00 to $2.50 per light for all material, including wire, lamp socket and lamp, but not fixtures or labor.

How to Install an Electric Light Plant

Although electric light outfits vary in design and details of construction, some general instructions for installation can be given which may be readily adapted to any plant.

Installation of Engine and Dynamo

After unpacking dynamo, it should be examined carefully to make sure there are no loose metal parts, such as a nail or nut in the frame of machine. The apparatus should also be cleaned thoroughly of any dirt.

Engine and dynamo should be mounted on a foundation of some sort, which if not purchased with outfit can be constructed from plans which will generally be furnished by the manufacturers of lighting outfits.

After dynamo is in place on foundation so that the driven pulley on dynamo shaft lines up with driving pulley on engine shaft, it should be bolted down.

Dynamo should then be belted up to gasoline engine with leather belting, being careful not to get it too tight.

Installation of Switchboard

Switchboard should be unpacked very carefully.

Switchboard panel should be mounted upright, taking care not to screw up nuts too tightly and break slate.

Panel should be located at point which permits wires to be most easily run between switchboard and storage battery. It should be perfectly plumb and securely fastened.

Wiring should then be accomplished in accordance with the instructions furnished with the particular make of outfit installed, and the current rate adjusted

All wiring should be well insulated and the different runs of wire should be as straight and pulled as tight as possible.

To make wiring connections with the various terminals, the insulation should be stripped from wire for about two inches at the end of wire, scraping until a bright, clean surface is obtained. The bared wire should then be bent tightly around the terminal stud and the nut tightened. In making connection at storage battery it is generally best to cover the exposed wire with tape and a heavy coat of grease or vaseline.

Installation of Storage Battery

In unpacking storage battery, examine glass jars carefully for breakage, as cracked jars must not be used.

Storage batteries must be located where reasonably dry and in a well ventilated room where the temperature is moderate.

As little exposed iron or metal work as possible should be in close proximity to storage battery, and any exposed metal there may be should be coated with asphaltum paint; also woodwork in the immediate vicinity.

Bright sunlight should not be permitted to fall upon a storage battery, which can be avoided by painting or whitewashing windows if necessary.

Jars should then be placed properly in accordance with directions accompanying the storage battery, and connections made. A final inspection should then be made to see that everything is all right.

After battery is set up and the cells filled with electrolyte liquid, and all connections made between dynamo, switchboard and storage battery, the outfit should be tested.

To do this, open all switches on the switchboard and start the gas engine by hand, with the switch on storage battery box closed to dry cell circuit, and when engine is up to speed turn rheostat to about mid-position. Voltage and polarity can be tested by a volt meter if one is at hand, or by connecting two short lengths of wiring to the dynamo armature terminals on the switchboard. Place the other ends of wire in a glass of water and add about a half teaspoonful of salt, without allowing the ends of wire to touch. When engine is up to speed, fine gas bubbles should come from the wires if everything is satisfactory. There will be very few bubbles around the positive wire and a great many around the negative wire.

CHAPTER XXXV

GASOLINE TRACTORS

The gasoline tractor, or self-moving unit, has become a powerful factor in farms of large and medium size.

The horse is limited in his capacity for work, but the use of the tractor is almost unlimited in capacity, speed and endurance. It has solved the labor problem, which has always been a serious problem on every large farm. It has

Drilling with a Gasoline Tractor.

been an influence in bringing under cultivation vast areas of our western prairies, and in all parts of the world it is making two bushels grow where one grew before. It has supplied man's greatest power need — that of tilling the soil, and cultivating and harvesting the resulting crop.

A tractor cuts down the amount of man labor required

on a big farm from a crew to the farmer and his boys or a man or two. It enables the farmer to do the work faster, not only enabling him to till a much larger acreage, but to do the work quickly, when the grain is ready to be harvested. It also assures more thorough tillage of the soil than is possible in any other way. These advantages combine to produce bigger and better crops and a bigger labor income. But, at the same time, the use of a tractor makes it necessary to buy more machinery, the cost of which, together with the cost of tractor itself, involves a

Threshing with a Gasoline Tractor.

large investment. On this account, careful figuring must be done to compute the cost of depreciation, repairs, upkeep and maintenance, as well as the interest on the large investment; and good business management must be employed in the use of traction equipment to make the increased income cover these overhead costs and still yield a net profit on the investment. For instance, if a tractor is employed three hours a day, it might show a loss, while if used six hours a day on an average, it would show a big profit. Or, if used on one operation at a time, it might show a loss, but if worked out to harrow, disc, pulverize and seed on the same operation, it would be very profitable.

12-25 H. P. Oil Tractor Engine.

The farm tractor is very adaptable and extensive in its application. Its usefulness commences before seeding time

Breaking, crushing and drilling virgin prairie into flax with a Gasoline Tractor.

Hauling wool with a Gas Tractor.

and does not end with the harvest. It can be used not only for plowing, discing, harrowing, harvesting, threshing, filling the silo, hauling the grain to market, etc., but can be

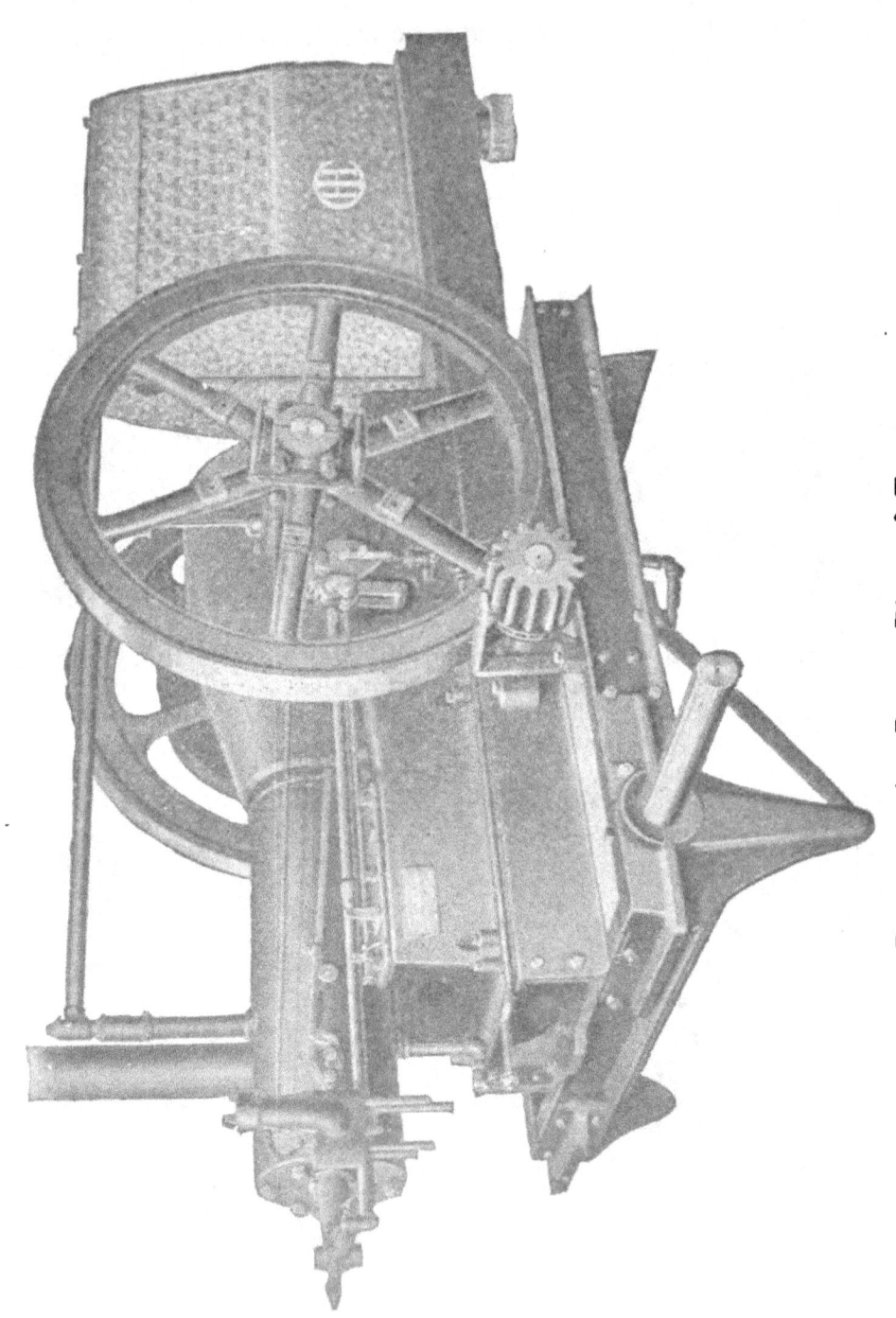

Representative Power Plant of Tractor.

used throughout the year for driving saw mills, road building, grinding and other heavy farm duties.

While it is not so long ago that it was thought the farm tractor was practicable only for plowing and threshing, and while it is probably needed for plowing more than for any other work on the farm, still on a large majority of farms its use is not profitable unless provision is made to keep it busy during as many months of the year as possible, on as great a variety of work as possible.

The owner of a tractor will also find that he can increase the usefulness of his outfit and add to his profit by doing custom work, such as plowing, seeding, harvesting, threshing, etc. A good manager can always make money at this, and, by gaining a reputation for good work, he will not have any trouble in getting all the outside work he can do.

The Power Plant

The power plant used in all gasoline tractors is an internal combustion motor, the fundamental operating principles of which are identical in all cases, and which are explained in the first few chapters of this book.

There is, however, a great variation in types, differences in the method of governing, ignition, reversing and cooling being particularly noticeable, the latter including air, water, oil and steam cooling devices.

The four cycle type of motor is universally used at present, there being no two cycle tractor motors that have risen far above the experimental stages. (Examination of two cycle and four cycle motors will be found in Chapter III.) Most of the smaller tractors are of the single cylinder type, which is the most economical of fuel, but not so steady in running as multiple cylinder types, owing to the longer interval between power strokes. Because of the limitations to the size of cylinders, this type must necessarily continue to be made in small units, the more powerful tractors now on the market using two, three and four cylinders. These are sometimes vertical and sometimes horizontal, and if horizontal may be either " twin " (side by side) or " opposed " (in line on opposite sides of the crank shaft).

The variation in engine speed is considerable, ranging

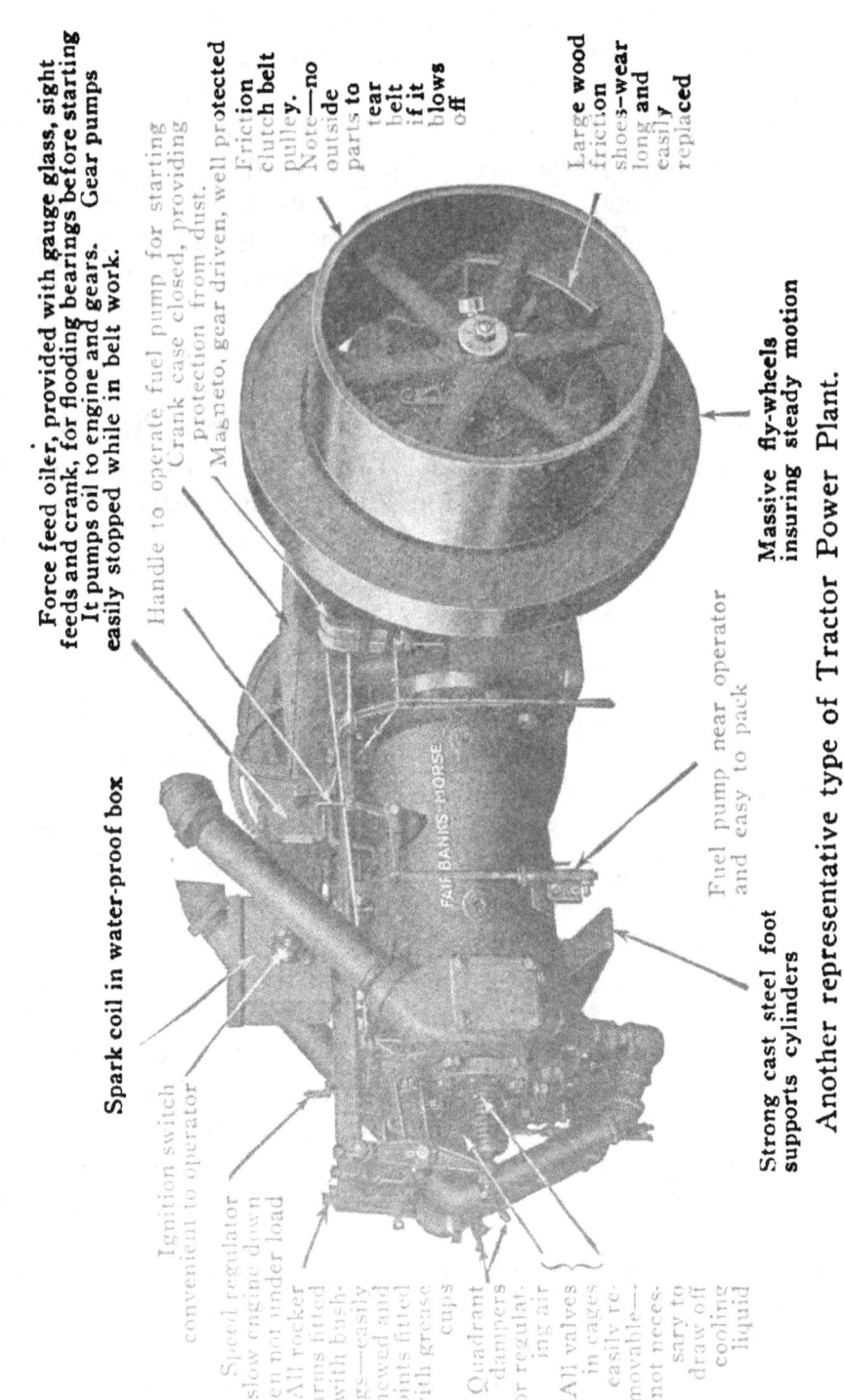

Force feed oiler, provided with gauge glass, sight feeds and crank, for flooding bearings before starting It pumps oil to engine and gears. Gear pumps easily stopped while in belt work.

Handle to operate fuel pump for starting

Crank case closed, providing protection from dust.

Magneto, gear driven, well protected

Friction clutch belt pulley. Note—no outside parts to tear belt if it blows off

Large wood friction shoes—wear long and easily replaced

Massive fly-wheels insuring steady motion

Spark coil in water-proof box

Ignition switch convenient to operator

Speed regulator to slow engine down when not under load

All rocker arms fitted with bushings—easily renewed and joints fitted with grease cups

Quadrant dampers for regulating air

All valves in cages easily removable—not necessary to draw off cooling liquid

Fuel pump near operator and easy to pack

Strong cast steel foot supports cylinders

Another representative type of Tractor Power Plant.

from 220 to 1,600 revolutions a minute, but as a rule not more than 550 revolutions are made at normal speed.

Fuels

Gasoline, kerosene and distillate are used as fuel in traction engines, opinion being divided as to the relative merits of kerosene and gasoline. Distillate, which is a low grade of kerosene, is largely used in the Pacific Coast States, and one or two other localities, but is not obtainable to any extent generally.

Kerosene is considered to be very successful where the engine is operating under a full and comparatively constant load, as in plowing, and of course is cheaper by the gallon, but it does not give as perfect combustion under varying conditions.

Generally a small amount of gasoline is used for starting and warming kerosene motors.

As a rule, the motor should be designed especially for the kind of fuel used, although a few types are equipped for burning either gasoline, kerosene or distillate. No engine, however, has been developed which will handle the different fuels equally well under all conditions.

The word " oil," as applied to fuel, is a rather indefinite term. Sometimes, and in some localities particularly, it is intended to mean kerosene; sometimes crude oil, unrefined; and in other cases it is understood to cover almost any refined product of crude oil.

The Frame

The motor, transmission gearing and rear axle are mounted on a frame or foundation, which keeps them always in the same relative position, regardless of the position tractor may be in.

In order to withstand the unusual strain or torsion in going over a rough or newly plowed field, in plowing up hill, crossing the irregularities of an old plowed field, etc., the frame must be of the greatest rigidity and strength.

Ordinarily some form of three-point suspension is employed to reduce strain to the minimum. The principle of three-point suspension is founded on the simple law of geometry that three points cannot be placed so that a flat surface cannot be made to pass through them.

In the tractor, the two rear drive wheel hubs are two of the three points, and the third point is in the front — either a single front wheel; or a two wheel construction which brings the front end of the frame into the shape of a " V " which rests in a single point on the front axle, usually having some form of ball and socket at this point.

The size of the frame is not the only indication of its strength, as the weight it must support and general design of the engine are important factors to be considered. For instance, the frame on a four cylinder engine may be made one-half the size of the frame of a clumsy one or two cylinder engine with heavy fly wheels and excessive vibration.

The general design, position of motor and gearing, etc., is also to be considered. Because of these factors, the various traction engine builders presumably make their frames of sufficient weight and strength to withstand strains and the amount of vibration of their particular design.

The Transmission

Practically all gasoline tractors are gear driven.

Ordinarily there are four general steps in the transmission of power from the motor to the draw bar.

The first is the transmission proper, which makes the first reduction in speed from the crank shaft.

The second step applies the power to the differential, the function of which is essential to assist in rounding corners, where one wheel must travel a longer distance than the other.

The third step delivers the power to the drive wheel rims.

The fourth step relates to the position of the drawbar and the relation of the balance of the engine to the hitch.

There are two types of gear transmission — the spur gear, in which the thrust is directly at right angles with each shaft and in but one direction; in other words, the two gears have a tendency to work apart — and the bevel gear, in which the thrust is no longer in one direction, but is divided between the shaft bearings and the thrust bearings behind each gear.

As a gas tractor works under such variable conditions, on rolling ground, and up grade as well as on level stretches, two speed and three speed transmissions are largely employed. The motor is, as a rule, run at medium speed a large portion of the time and the load behind adjusted. For instance, when a tough spot is reached the operator drops back into low speed. This increases the draw bar pull possibly 40%, allowing the engine to go through the emergency without jerking or overstrain. Then again, there are certain kinds of work where it is of value to run at a speed faster than the plowing speed, as for example, returning from market with empty grain wagons, where it is very desirable to make time on the road. In a three speed transmission, the high speed is desirable here.

The gearing is a most important factor and should be not only of ample proportions and of a material that will reduce to a minimum the possibility of wear or breakage, but the gears should be cut with perfect accuracy.

The hauling power of a tractor depends upon the speed it is geared to travel. One tractor may haul as many plows as another and yet do much less work if geared to run slower. Too high a speed, however, is very hard on the machine, and will soon rack it to pieces. Although it would be easily possible to gear a tractor to run five, six or ten miles an hour on the road, it would be very bad judgment, and possibly be the cause of ruining the outfit the first year. Three or four miles an hour is as fast as any tractor or farm machines should be driven over the road.

The backing speed of a tractor is generally about one-half its forward speed.

A system of speed regulation is generally employed, whereby a variation in speed of from 25 to 40 per cent. is permitted with the same arrangement of gearing.

Wheels

There is a great deal of variation as to the height and width of tractor wheels.

One factor, however, is always essential. That is great

strength to meet the vigorous demands of plowing, as well as all requirements of traveling on the road.

Generally speaking, the larger the wheels the better, as it takes less power to pull a load over soft or loose ground on large wheels than on small ones. At the same time, however, wheels must not be cumbersome, clumsy or unsafe.

Wheels are generally of two types, cast and built up.

Built up wheels are also of two general types. In one

Typical Oil Tractor for soft ground work.

flat spokes are used, the heads of spokes being upset or flattened out and pressed against the rim of the driver. The spokes are coupled to the hub by rivets, and the weight of the engine is carried by pressure against the rim of the wheel, which, in turn, is carried by the rivets. In the other type, which is known as the suspended type, the spokes have round heads and pass completely through the rims. The weight of the engine is suspended or held up by the tension of the spokes.

In order to make the use of large wheels practicable, as

much driving strain as possible must be removed from the spokes, which is ordinarily done by connecting the rim of the master gear with the rim of the drive wheel. The rims should be of sufficient weight to withstand hard service.

Plowing

Of all uses of power, there is no one which surpasses in importance that of plowing.

Practically all traction plowing is now done with specially designed gang plows. Two types are in most general use — the disc and mold-board types.

Tractor with curtains for protection of working parts.

As great variation is presented in the design and construction of gang plows as in horse plows, variations in shapes that have been worked out to meet local conditions and differences of opinion, applying just the same in gang plows.

A tractor working 24 hours a day with 4 men, 2 each 12 hours, will plow 40 to 60 acres a day under average conditions. As plowing should be done quickly when the ground is just right, the advantage of a tractor is easily apparent. The tractor also permits deeper plowing which is an advantage, especially in localities where it is important to liberate the fertility of new soil. The use of a tractor insures thorough plowing, and as the difference between good plowing and bad plowing is often the difference between success and failure, the use of a tractor may be regarded as crop insurance.

Disc Plows

Disc plows are popular with gas tractioneers as they are more easily manipulated and cover more ground in a given time than the mold-board type. They accommodate them-

Gas Plowing.

selves to uneven surfaces, tend to roll over obstructions and permit plowing a continuous furrow around field without the labor and loss of time in lifting plows at corners or headlands, or plowing corners with horses.

Disc plows are not so expensive as mold-board plows, but are not generally considered as efficient. They also present greater difficulties in providing a well balanced hitch, small gangs being somewhat difficult to turn corners and large gangs being heavy to lift, unwieldy and hard to keep in line. Medium sized gangs of five, six and seven discs, with a load of ten to fifteen discs well balanced, give very excellent results.

Mold-Board Plows

Mold-board plows, which are generally considered to do the best work, consist of several gangs of plows hung on a single frame, each gang being lifted and dropped either by power or hand. The arrangement of plowshares into bottoms independent of each other permits surface irregularities to be met and in case of a solid obstruction, one or two bottoms may be released without damage to the entire outfit. Ordinarily about six to eight bottoms are regarded as a fair load for the gasoline tractor.

Harrowing, Seeding, Etc.

A gas tractor will haul as many implements as a dozen horses, and upwards. Plow, seeder, packer, harrow, in fact the entire ground working outfit can be placed behind tractor. Say a tractor hauls four 10 feet seeders and two 20 feet harrows and covers 7 to 8 acres per hour, compare the cost of accomplishing this work with what two men could drill and drag it for. The tractor will enable you to prepare a better seed bed and do a better job of seeding and leave the ground in better shape, at much less cost than animal power.

Harvesting

With a tractor, several binders can be used at once, at faster speed and each binder doing more work than would be possible to do with horses. It is not at all unusual to see a tractor hauling five or six 8 ft. binders, making a cut nearly 50 ft. wide. This can be kept up 24 hours a day if desired by having two shifts of men. The progress with such an outfit is readily apparent.

Threshing

In addition to traction work, the farm tractor, by means of a belt drive, can be used to run thresher, including separator, feeder, wind stacker, weigher, etc. It is superior to a steam threshing outfit in many ways — gasoline or kerosene is easily obtainable and easily carried; no expense for teams to haul coal and water. There are no sparks to set

Cab of one of the leading modern Tractors.

fires — engine can be set among stacks where it would not be safe to use steam. There is no hole to dig in ground when standing on a hill, as it will run with either end up as well as on the level. No loss of time to raise steam; no getting up at four o'clock to build a fire — not necessary to go four or five miles out of the way to avoid bridges over which it would not be safe to take a steam outfit; can be operated by one man and does not require a special licensed engineer.

Other Belt Driven Operations

Besides threshing, a tractor can be used to operate any farm machinery by means of a belt.

It will run corn sheller, corn husker, shredder, feed grinder, ensilage cutter, wood saw, well driller, rock crusher, concrete mixer, pump and many other machines. As a matter of fact, however, the fuel cost is too great for it to be profitable in driving very light machinery on long runs, and for any amount of this work it pays to get a small horse power stationary or portable engine to be used for this purpose.

Hauling

Any kind of hauling can be done by a tractor. If desired, from ten to twenty wagons can be hauled in tandem for taking grain to market. With but one man, a tractor can also be made to pay a good profit in helping to build roads — in grading and hauling surfacing material, as well as in hauling lumber, logs, etc.

Hitches

A well designed hitch will increase the efficiency of the tractor and all of the machines hauled behind it, improve the quality of work done, and decrease the amount of wear and tear of the machine.

Among the qualifications of a good hitch are the following: It must be convenient and so designed that it will be a simple matter to attach to or disconnect the tractor and machines from it. It should be easy of manipulation.

Modern Oil Tank Wagon.

Durability is also essential as the strain resulting from hauling a number of heavy machines behind an engine is naturally excessive. At the same time, however, it should be as light as possible, so as not to be a burden to pull. It should be designed with a view of being easily adapted for handling so many different machines and working under as many different conditions as possible. It should also be flexible so that it will readily conform to the surface of the ground, and so connected that hitch, tractor or machines being hauled will not be subjected to any undue strains when turning.

Most tractor builders will offer advice and recommendations on this subject, and many of them will furnish detailed plans showing just how hitches can be constructed by a handy man or blacksmith from timbers, poles, etc., usually obtainable in any small town. As different farmers, however, naturally use different methods in soil culture, it is largely a matter of each farmer designing hitches to meet his own ideas and requirements.

Brake Horse Power

This is the actual power which the engine will develop at the pulley for belt work.

Draw Bar Horse Power

This is the power the engine will develop at the draw bar for pulling a load, and is equal to the pounds exerted times the feet traveled per minute divided by 33,000. For example, if engine is exerting draw bar pull of 3.75 pounds, at the rate of 3 miles per hour, the feet traveled per minute would be $3 \times 5{,}280 = 15{,}840$. This divided by $60 = 264$ feet per minute, and draw bar horse power is equal to $264 \times 3{,}750 = 990{,}000$. This divided by 33,000 $= 30$ H. P.

The Care of a Tractor

Good care is of greater importance with a tractor than with any other piece of farm machinery for the reason that a larger investment of money is involved. Care is just as important to the profit a tractor earns as output.

Depreciation depends entirely upon the care taken of the machine. Some farmers will have as much depreciation in a single year as others will in three years. This is a matter that is individually up to the owner of a tractor.

Good, steady work, without any attempt at unusual feats or taking chances at overstraining is one of the first essentials in keeping a tractor in good condition.

The next step is to spend an hour a day, at some certain time, going over the engine to make sure it is in good condition — looking over the mechanism carefully to see that all nuts are tight and that there are no broken parts, examining bearings, etc., as well as filling the supply tanks, lubricators, grease cups, etc. Particular care should be taken to seeing that all bearing surfaces are running cool. Connecting rod bearings should also be tested, and if loose should be taken up at once. It is a good plan to do this work first thing in the morning and have an extra hand to help with it, so that the outfit can be started at actual work as early in the day as possible.

More thorough inspection should be made once a week at which time exhaust valves should be ground in (inlet valves do not require grinding). Directions for grinding valves will be found in Chapter XI on Engine Trouble. The crank case and cylinder oil system should also be drained, flushed out with gasoline or kerosene, and replaced with good, clean oil once a week.

Overhauling and Winter Housing

The majority of engines have a more or less idle spell of three or four months in the winter.

A great deal of trouble and delay during the rush of the next busy season when time counts, can be avoided by giving the equipment a general overhauling at the end of the season. This is a much better time to do it than at the beginning of next season, as the operator knows just where the trouble is, what needs replacement, etc., and is much less liable to overlook anything needing attention when these facts are fresh in his memory.

All worn and broken parts should be repaired or replaced. Pistons should be examined for broken rings, bearings adjusted, re-babbitted, etc. General instructions on the overhauling and repair of an internal combustion motor, Chapter XII; also instructions for re-babbitting in Chapter XXI on " Pulleys, Shaftings, and Bearings " will be found useful in connection with this work.

If engine is to remain idle during a large part of the winter, it should be protected from the cold and dampness.

Exposure to cold weather will cause moisture to condense on all parts of the outfit, inside and out.

Engine should be kept under cover when not in use.

Magneto should be removed and placed in a dry place; it is a good plan to take it in the house.

About a quart of thick oil should be poured into each motor cylinder, and crank turned over until oil has been spread over the walls of cylinder. Be sure all priming and relief cocks are closed. Drain all water from cylinder.

Remove valve caps, pour thick oil on valves and see that it is thoroughly worked between valve and seat; grease valve caps and put back.

Cover such parts as carburetor, governor, fan belt, etc., with canvas.

Be sure all water has been drained not only from cylinders, but from water pump, piping and all connections.

Before starting engine again, clean the old thick oil out by flushing cylinders and gear cases with kerosene; prime the cylinders with a generous quantity of good cylinder oil before starting motor and fill the gear cases with good quality transmission grease.

What a Tractor Will Do

It can be run twenty-four hours a day in case of rush work.

It costs nothing when not working.

It enables harvesting at exactly the time crop is in best condition.

It permits early plowing, when it is generally too hot for horses.

It makes it possible to pull plows right behind the binder, if desired, so that ground will not be left to dry out.

It will lighten your wife's work, as well as your own, as she will not have to feed a gang of hungry men.

It makes it unnecessary to keep men to take care of horses, pay them wages, feed them, etc.

It can be run day and night in hottest weather, when the heat compels horses to be rested often.

It can be used for baling hay and will haul it in three trips instead of ten.

Daily Records

The importance of keeping track of results when large investments are made, as in the purchase of traction equipment, demands a daily report of cost and results.

This should show for each day of the year the number of miles traveled and acres plowed; the total amount of fuel and lubricating oil used; the cost of repair parts and length of time required to make repairs; time spent in inspecting and adjusting equipment, and all other charges and incidental expenses. This will enable the farmer to *know* just how much the tractor is saving him on his work, and just how profitable it is on the custom work he does with it.

The Tractor on the Road

(The following sensible advice from an old experienced driver of steam tractors is so valuable for the amateur that we reproduce it entire. In general its information is applicable to gasoline tractor handling as well as to the manipulation of the steam traction engine.)

It is something of a trick to handle a traction engine on the road. The novice is almost certain to run it into a ditch the first thing, or get stuck on a hill, or in a sand patch or a mudhole. Some attention must therefore be paid to handling a traction engine on the road.

In the first place, never pull the throttle open with a jerk, nor put down the reverse lever with a snap. Handle your engine deliberately and thoughtfully, knowing beforehand just what you wish to do and how you will do it. A traction engine is much like an ox; try to goad it on too fast and it will stop and turn around on you. It does its best work when moving slowly and steadily, and seldom is anything gained by rushing.

The first thing for an engineer to learn is to handle his

throttle. When an engine is doing work the throttle should be wide open; but on the road, or in turning, backing, etc., the engineer's hand must be on the throttle all the time and he must exercise a nice judgment as to just how much steam the engine will need to do a certain amount of work. This the novice will find out best by opening the throttle slowly, taking all the time he needs, and never allowing any one to hurry him.

As an engineer learns the throttle, he gradually comes to have confidence in it. As it were, he feels the pulse of the animal and never makes a mistake. Such an engineer always has power to spare, and never wastes any power. He finds that a little is often much better than too much.

The next thing to learn is the steering wheel. It has tricks of its own, which one must learn by practice. Most young engineers turn the wheel altogether too much. If you let your engine run slowly you will have time to turn the wheel slowly, and accomplish just what you want to do. If you hurry you will probably have to do your work all over again, and so lose much more time in the end than if you didn't hurry.

Always keep your eyes on the front wheels of the engine, and do not turn around to see how your load is coming on. Your load will take care of itself if you manage the front wheels all right, for they determine where you are to go.

In making a hard turn, especially, go slow. Then you will run no chance of losing control of your engine, and you can see that neither you nor your load gets into a ditch.

GETTING INTO A HOLE

You are sure sooner or later to get into a hole in the road, for a traction engine is so heavy it is sure to find any soft spot in the road there may be.

As to getting out of a hole, observe in the first place that you must use your best judgment.

First, never let the drive wheels turn round without doing any work. The more they spin round without helping you, the worse it will be for you.

Your first thought must be to give the drive wheels something they can climb on, something they can stick to. A heavy chain is perhaps the very best thing you can put under them. But usually on the road you have no chain handy. In that case, you must do what you can. Old hay or straw will help you; and so will old rails or any old timber.

Spend your time trying to give your wheels something to hold to, rather than trying to pull out. When the wheels are all right, the engine will go on its way without any trouble whatever. And do not half do your work of fixing the wheels before you try to start. See that both wheels are secure before you put on a pound of steam. Make sure of this the first time you try, and you will save time in the end. If you fix one wheel and don't fix the other, you will probably spoil the first wheel by starting before the other is ready.

Should you be where your engine will not turn, then you are stuck indeed. You must lighten your load or dig a way out.

BAD BRIDGES

A traction engine is so heavy that the greatest care must be exercised in crossing bridges. If a bridge floor is worn, if you see rotten planks in it, or liability of holes, don't pull on to that bridge without taking precautions.

The best precaution is to carry with you a couple of planks sixteen feet long, three inches thick in the middle, tapering to two inches at the ends; also a couple of planks eight feet long and two inches thick, the latter for culverts and to help out on long bridges.

Before pulling on to a bad looking bridge, lay down your planks, one for each pair of wheels of the engine to run on. Be exceedingly careful not to let the engine drop off the edge of these planks on the way over, or pass over the ends on to the floor of the bridge. If one pair of planks is too short, use your second pair.

Another precaution which it is wise to take is to carry fifty feet of good, stout hemp rope, and when you come to a shaky bridge, attach your separator to the engine by this rope at full length, so that the engine will have crossed the bridge before the weight of the separator comes upon it.

Cross a bad bridge very slowly. Nothing will be gained by hurrying. There should especially be no sudden jerks or starts.

SAND PATCHES

A sandy road is an exceedingly hard road to pull a load over.

In the first place, don't hurry over sand. If you do you are liable to break the footing of the wheels, and then you are gone.

In the second place, keep your engine as steady and straight as possible, so that both wheels will always have an equal and even bearing. They are less liable to slip if you do. It is useless to try to " wiggle " over a sand patch. Slow, steady, and even is the rule.

If your wheels slip in sand, a bundle of straw or hay, especially old hay, will be about the best thing to give them a footing.

HILLS

In climbing hills take the same advice we have given you all along: Go slow. Nothing is gained by rushing at a hill with a steam engine. Such an engine works best when its force is applied steadily and evenly, a little at a time.

If you have a friction clutch, as you probably will have, you should be sure it is in good working order before you attempt to climb hills. It should be adjusted to a nicety. When you come to a bad hill it would probably be well to put in the tight gear pin; or use it altogether in a hilly country.

When the friction clutch first came into use, salesmen and others used to make the following recommendation (a recommendation which we will say right here is bad). They said, when you come to an obstacle in the road that you can't very well get your engine over, throw off your friction clutch from the road wheels, let your engine get under good headway running free, and then suddenly put on the friction clutch and jerk yourself over the obstacle.

Now this is no doubt one way to get over an obstacle; but no good engineer would take his chances of spoiling his engine by doing any such thing with it. Some part of it would be badly strained by such a procedure; and if this were done regularly all through a season, an engine would be worth very little at the end of the season.

Threshing Rig Advice

(The tractor is so generally used for hauling and operating a threshing rig nowadays that a few words of advice regarding the thresher part of the outfit will not come amiss. This does not pretend to exhaust the subject by any means, but the few hints are worthy of pondering.)

Before you uncouple your traction engine go over the threshing place carefully; make note of the slope of the ground and the direction of the wind. The best position you can secure relatively to the wind is when the straw on the road to the stack moves a little to one side but in the same general direction as the wind.

See that you have room for your stack, that loaded and unloaded wagons can pass, and that nothing can interfere with your belts. Bearing these points in mind, select the position that will be most convenient and haul your thresher in place. If the ground is not level, dig holes for the wheels which are too high and block the rear wheels well. It is best to carry a level with you, as this saves time in the end. Be careful that the machine is level crosswise. It is not so necessary that it should be level lengthwise. In fact, it is of some advantage to have the cylinder end four to six inches higher than the cleaner end; but remember it must not lean the other way. See that the blocks of the right hind wheel are tight, so that the pulling of the belt will not disturb the setting. If there are jack-screws above the front and rear axles, to make the four corners solid adjust them and screw the nuts down tight. Now you can uncouple and pull your traction engine to such a position that the driving-wheel and the pulley on the thresher are in line. Take your time and do this right. It is better to spend a little more time now than to have your belt run off when you are running with a heavy load on the machine. Most of the pulleys are crowned, that is, they are a little larger in diameter in the center than at the sides, and the tendency of the belt is to run in the center; but if the machines are not in line and the shafts not horizontal, the belt will nevertheless run off. Before you put the belt on, go over the machine carefully and clean all the bearings and oil holes well. Take off the belt-tightener pulley and clean the oil chambers and the spindle, oil them and put them back. If the machine is new, more care must be used when doing this, as paint may have got into the bearings and oil holes, in which case it is best to remove the shaft and carefully scrape the paint off. Wipe out the bearing and the oil hole with clean waste, oil them well, and replace the shaft. After you are sure they are all cleaned, put a few drops of oil in each oil hole. Use only No. 1 machine oil with good body. If the machine uses grease, see that it is good. Don't use axle grease, as it very often contains resin, which will deposit on the bearings and cause them to heat.

Now put on the main belt, and be sure that the machine runs the right way. You can easily see if the cylinder will pull the straw in or not. If it should run the wrong way, cross your belt. It is not good practice to run with a crossed belt, as it takes more power. Sometimes, if the belt is short, and the difference in diameter between the pulleys is great, a crossed belt will give more power, but it is better not to run it crossed except when it is necessary.

GLOSSARY
OF GASOLINE ENGINE TERMS

ADVANCE SPARK — A spark taking place just before the time piston is at its highest point.

AIR COOLED — Engines in which cooling is accomplished by air.

BABBITT — A special metal used for lining bearings.

BATTERIES — (Cells) — Equipment for generating electric current.

BACK FIRE — An explosion which occurs in the carburetor, base or muffler instead of in combustion chamber.

BELT DRIVE — Transmission of power by means of belt running from a pulley on the engine to a pulley on driven machine or to a line shaft.

BORE — Inside diameter of cylinder.

BEARING — Any surface supporting a revolving shaft of any kind.

BUSHING — A replaceable metal lining or sleeve for bearing.

B. H. P. — Brake horse power. Determined by brake test.

CAM — Device for timing or operating valves.

CAM SHAFT — Shaft which turns cam.

CARBURETOR — Device which mixes liquid fuel with air to form explosive mixture.

CHARGE — A cylinderful of explosive mixture.

COCK — A spout for drawing off liquid, air, etc.

COMBUSTION CHAMBER — Portion of cylinder where mixture is compressed and fired.

COMPRESSION COCK — A spout for releasing compression.

COMMUTATOR — Part of timer which breaks the current.

COMPRESSION — Pressure in combustion chamber.

CONNECTING ROD — Forging connecting piston to crank shaft.

CRANK — To turn fly wheel of motor for starting.

CRANK SHAFT — Shaft receiving the back and forth motion of piston and transmitting it in rotating form to where needed.

CRANK PIN — Part of crank shaft that connecting rod is fastened to.

CRANK CASE — Receptacle which encloses the crank shaft.

CYCLE — One-half revolution of the crank shaft.

CYLINDER — Part of engine in which piston moves.

CYLINDER HEAD — Top part of cylinder.

DRY BATTERIES — A type of battery.

DIRECT CONNECTED — Engine directly connected to a single machine to be operated.

DISTILLATE — A low grade of kerosene used largely as engine fuel on the Pacific coast.

DYNAMO — Form of electric current generator differing from magneto in that part of current is utilized to magnetize the field.

EXHAUST — The discharge of burnt mixture after ignition.

FLOAT — Part of carburetor to regulate fuel level.

FLOODING — Flooding the carburetor with too much gasoline.

FLOAT FEED — A type of carburetor in which the gasoline level is automatically maintained by means of a cork float.

FLOOD CHAMBER — Part of carburetor which liquid gasoline first enters.

FOUR CYCLE — Type of engine in which each complete power impulse is produced with four strokes of the piston and two revolutions of the crank shaft.

FUEL PIPE LINE — Piping through which fuel runs from supply tank to engine.

FORCE FEED — Type of lubrication in which lubricant is forced through separate pipes to bearings.

FUEL TANK — Receptacle for holding fuel supply.

FLY WHEEL — Heavy wheel to steady running of engine.

GASKET — A piece of packing to insure a tight joint.

GEAR — A toothed wheel.

GOVERNOR — Device for controlling or governing speed of engine.

GREASE CUPS — Lubricant containers located directly over bearings.

HEADER or MANIFOLD — Piping or casting through which the exhaust and intake travels.

HIGH TENSION — Electric current at high voltage.

HORIZONTAL ENGINE — Engine in which cylinders are horizontal.

"HIT-AND-MISS" — Type of governor which causes engine to miss several explosions for every one that ignites.

HOPPER — Receptacle in which cooling water is contained in hopper-cooled type of engines.

IGNITER PLUG — Device where spark occurs in make and break ignition.

JUMP SPARK — A system of engine ignition.

LIFT ROD —(Tappet)— Rod operated by cam to open valves.
LOW TENSION — Electric current at low voltage.

MAGNETO — Device for generating current.
MAKE AND BREAK — A system of engine ignition.
MIXTURE — Liquid gasoline mixed with air to form explosive gas.
MISFIRE — Missing explosions.
MUFFLER — Device for muffling or silencing exhaust.

NEEDLE VALVE — Small valve regulating flow of fuel or oil.

PISTON — Part of engine which moves back and forth in cylinder.
PISTON PIN —(Wrist Pin)— Metal pin fastening piston to connecting rod.
PISTON RING — Split cast iron ring used on piston.
PORT — Opening for intake or exhaust of mixture.
PRIMING CUP — Small cup usually placed on cylinder, for priming with fuel to start motor.
PORTABLE ENGINE — Engine designed especially to permit of being readily moved from place to place.
POSITIVE PRESSURE — Method of lubrication in which oil is forced to working parts by compression of the engine.
PRE-IGNITION — Ignition of charge before the proper time.

RETARDING SPARK — A spark taking place after piston has reached highest point.
R. P. M.— Revolutions per minute.

SPARK COIL — Device which receives current from the source of supply and transmits it to point of ignition within combustion chamber.
SKIDS — Usually some sort of wooden runners on which engine is mounted and dragged over the ground.
SPARK PLUG — Device where spark occurs on jump spark ignition system.
STRIP — Tear down or dismantle.
STROKE — Distance traveled by piston.
STATIONARY ENGINE — Type of engine intended particularly for setting stationary in one place, usually mounted on a sub base.
SIGHT FEED — Lubrications in which feed of lubricant is in sight.
SLIDE CONTACT — A form of timer.
STORAGE BATTERY — Type of battery which stores current for future use.
SPLASH SYSTEM — System of lubrication by splashing oil onto crank shaft.

THROTTLE — Device for controlling speed of engine.
TIMER — Device for controlling time of spark.

TIMING GEARS — Gears for driving timer and cam shaft.

TRACTOR — Self-moving engine, used largely for plowing, hauling, etc.

THERMO-SYPHON — A system of water cooling by gravity.

THREE-PORT — Type of two-cycle engine in which three ports or openings are made use of in the process of taking mixture into the cylinder and expelling it.

TWO-PORT — Type of two cycle engine in which two ports or openings are made use of in the process of taking mixture into cylinder and expelling it.

TWO CYCLE — Type of engine in which each complete power impulse is produced with two strokes of the piston and one revolution of the crank shaft.

VALVE — A movable cover, lid or partition for opening and closing communication in a tube or orifice.

VERTICAL ENGINE — Engine in which cylinders are vertical.

WATER COOLED — Engines in which cooling is accomplished by water.

WATER JACKET — Hollow space in cylinder casting of water cooled engines surrounding combustion chamber through which cooling water circulates.

WET CELLS — A type of battery.

INDEX

www.ingramcontent.com/pod-product-compliance
Lightning Source LLC
Chambersburg PA
CBHW081118170526
45165CB00008B/2479

* 9 7 8 1 5 4 8 1 5 3 1 9 9 *